Earth Science
Discovering the secrets of the earth

ROCKS

Atlantic Europe Publishing

Atlantic Europe Publishing

First published in 2000 by
Atlantic Europe Publishing Company Ltd

Reprinted 2001

Author
Brian Knapp, BSc, PhD

Art Director
Duncan McCrae, BSc

Editors
Mary Sanders, BSc and Gillian Gatehouse

Illustrations
David Woodroffe and Julian Baker

Designed and produced by
EARTHSCAPE EDITIONS

Reproduced in Malaysia by
Global Colour

Printed in Hong Kong by
Wing King Tong Company Ltd

Suggested cataloguing location
Knapp, Brian
Earth Sciences set of 8 volumes
Volume 2: *Rocks*
1. Geology – Juvenile Literature
2. Geography – Juvenile Literature
550
ISBN 1 862140 38 3

Picture credits
All photographs are from the Earthscape Editions photolibrary except the following:
(c=centre t=top b=bottom l=left r=right)
Keith Ronnholm 18bl; *USGS* 14b, 17tr.

This product is manufactured from sustainable managed forests. For every tree cut down, at least one more is planted.

Contents

Chapter 1: The nature of rocks

ROCKS are the solid materials that make up the earth's CRUST. They make up the mountains and the hills, and have been eroded to leave valleys and plains.

Rocks catch our eye when they form dramatic or curious shapes, such as tall cliffs and pillars, arches, and natural bridges, but we soon become aware that the rounded rocks or stepped cliffs are made of materials of various colours and hardness – that is, the landscape is, in large measure, shaped from rocks.

(Left and above) The rounded forms of this landscape are due to the effect of the weather on granite rocks. JOINTS were produced as land above this former mountain core was worn away, allowing the rock to expand. The uniform crystals in the granite allowed the rock to become weathered into balls.

(Right and below) The staircase of this landscape has been produced from alternating bands of hard and soft rock. Hard rocks, like sandstone and limestone, make the small cliffs, or risers in the staircase, while softer shale rocks make the treads. All such rocks contain fossils and tell of a time when their materials were deposited in a sea. They are the remains of even older landscapes.

The pictures below show a detail of grains of sandstone and an ammonite fossil, both indicators of sedimentary rock.

The properties of rock

The nature of a rock becomes clearer if you hold a piece in your hand. It is not a single material (which would make it a **MINERAL**), but a collection of small pieces, sometimes **CRYSTALS**, and sometimes **GRAINS**, that have either grown together, become fused together, or been cemented together.

These crystals and grains help give a rock its colour, hardness, weight, texture, and resistance to the weather.

Usually a rock is made up of many grains of about the same size, although the grains may not be made of the same minerals.

So, when we talk about rocks, we are more interested in the way the grains or crystals behave than in the minerals themselves, unless the properties of the minerals give special properties to a rock. For example, if a rock is made of salt, we are interested in its

minerals because they readily dissolve in water.

There is not much difference between the resistance of many minerals, which is one reason why most volcanic rocks are hard and stand up to the weather.

Some rocks (called IGNEOUS ROCKS and METAMORPHIC ROCKS) are made up of tightly interlocking crystals that give them their properties. Others (called SEDIMENTARY ROCKS) are made of grains that have been cemented together. Some cements are very strong; others are weak. Their strength affects the way the rock can stand up to attack by the weather. So, for sedimentary rocks it may be of more interest to know about the way the cement behaves than about the particles themselves.

(Above) Pink granite and dark folded rocks that once formed deep inside a mountain range.

Rock cycles

All rocks on the earth's surface are made of second-hand materials. No rocks remain that were formed at the start of the earth's history. There are two reasons for this. In the first place, the earth's crust is split up into a number of large slabs, called TECTONIC PLATES, which continually move around. When they split

(Below) This diagram shows the rock cycle. New rock is formed in the ocean and also by magma rising from under the continent. Mountains, formed by plates colliding, are eroded by the weather and rivers to produce fine particles that are carried to the sea and deposited in layers. Here, they are compressed into new rock. When continents collide, they will be folded into new mountains before being eroded again. Thus the cycle continues.

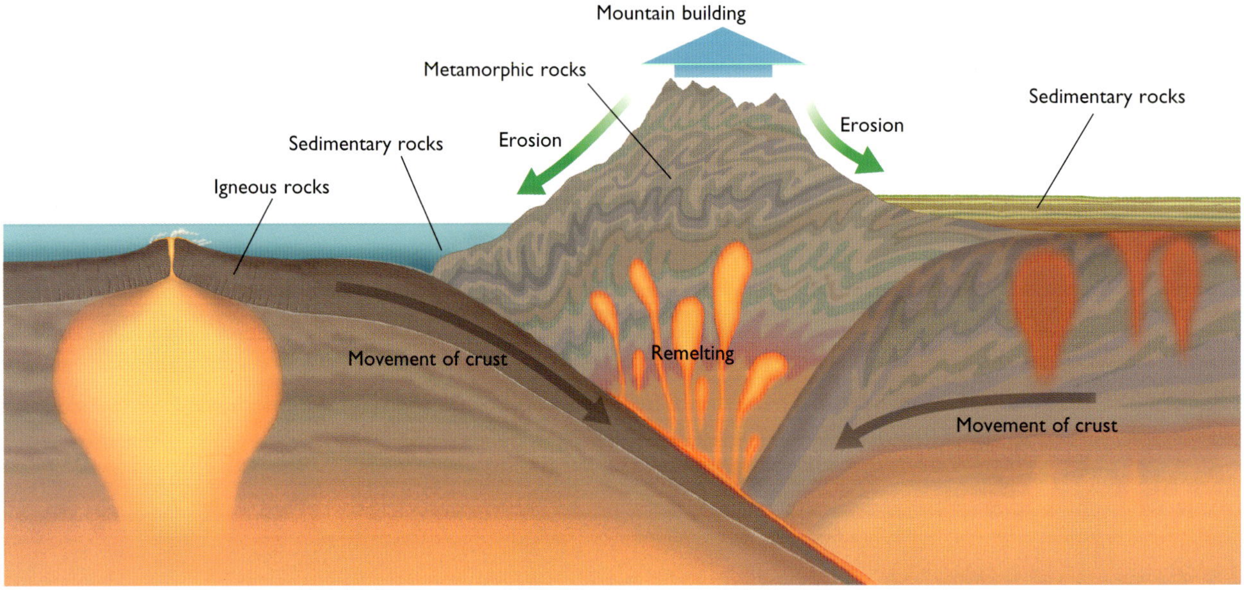

apart, they allow new molten rock to reach the surface. All of the ocean floors are made of relatively new rocks because of this movement. When they collide, tectonic plates are often pushed back down into the earth. Much old rock is lost in this way. (For more information on tectonic plates, see the book *Plate Tectonics* in the *Earth Science* set.)

But even ancient rocks that have remained on the surface of the earth for billions of years were attacked long ago by the forces of the weather, broken down into small fragments, and carried to the seas, where they were laid down in layers. Some materials were also carried in solution. In the sea, the actions of chemicals and living things have played a further role, forming new rocks, such as limestone, while the burial of living things on land and sea has produced rocks, such as coal.

(Above) The layered nature of these rocks is clearly seen through the colours. This is a muddy rock called shale.

Few of these rocks were left to rest for long, however, for the movement of the tectonic plates eventually made much of the ocean into land, and most of the sediments into new mountain ranges, or at least into high tablelands.

So, the materials of rocks are continually recycling. New molten materials from within the earth become parts of mountains that are worn down into sediments. Sediments become crushed into mountains and made into parts of tectonic plates, and parts of tectonic plates eventually get pushed back inside the earth. This kind of recycling has being going on ever since the crust began to form at least 4 billion years ago.

But, at the same time, each step from liquid rock through uplift, erosion and new rock formation is a chance occurrence, because uplift will never be the same twice, and rock that was formed from rain-washed mountains in one turn of the cycle might be eroded in desert conditions next time around. It is both recycling and random differences in the climate and uplift that produce the amazing variety in the thickness and character of the rocks that we find today on the earth's surface.

Rock-forming environments

Rocks are to a geologist as the scene of a crime is to a detective. The samples that a geologist collects are full of clues to their past. By looking at a sample, it is possible to discover a lot about how the rock got to the place where it was found, how it formed, and the origin of the materials that formed it, whether it was

(Below) A road cutting provides a wonderful window on the world's rocks. Here you can see the rocks, layer upon layer. They stand out because of differences in colour. These were once laid down under the sea. Now they form a mountain summit. As part of the mountain system they were folded, and here we can see just one part of the fold, a downward bend called a syncline. And even before the cutting was made, much of the rock was eroded away, for the upstanding limbs of the syncline now end in the air.

heated or squashed, and so on. In this way it is possible to trace some rocks back for billions of years. The conditions under which the rock formed are called the rock-forming environment.

Classes of rock

One of the remarkable conclusions from the study of rocks is that nearly all rocks on the surface today are made from fragments of other rocks. The result is, that if you were to crush a rock and give it to a chemist, the chemist would tell you that the overall composition was much the same as any other sample. (The main exceptions to this are rocks like coal and limestone.) So we do not necessarily look for the distinctive minerals in a rock – for they may be quite similar to others – but for the last process that the rocks went through. This is the reason why rocks are often classified by their origin and texture, rather than by their chemical composition.

Geologists, therefore, divide up rocks into three basic classes:

- igneous rocks – those that have formed from molten material within the earth;
- sedimentary rocks – those that have formed from fragments of other rocks; and
- metamorphic rocks – those that have been considerably altered by heat or pressure since they were formed.

Broadly, when looking at rocks of the three classes, you will find that:

- igneous rocks have randomly arranged interlocking crystals;
- sedimentary rocks are made of grains cemented together; and
- metamorphic rocks contain crystals that are mainly organised into bands (sometimes giving the impression that the rock has been 'smeared out').

The nature of each class of rock will be described in the following chapters.

Chapter 2: Rock-forming minerals

Although the chemical elements occur in many thousands of minerals, just a few minerals are common in rocks. They are called the rock-forming minerals.

It is important to know about the minerals that form rocks because their hardness, solubility, and other properties can help explain why the rock is hard or soft, and how it was formed. A brief description of the rock-forming minerals is given below. For more detail see the book *Minerals* in the *Earth Science* set.

Feldspars

Feldspars (named after the German for 'field crystals') are the most common minerals found in the earth's crust. About half of all the earth's crustal rocks are made from feldspars. There are two kinds. One kind (orthoclase feldspar) is white

(Below) This piece of granite is made up of several different minerals. The pink ones are plagioclase feldspar, the white ones are orthoclase feldspar, the large black minerals are hornblende, the small black one biotite, and the glassy grey minerals are quartz.

or grey and the other kind (plagioclase feldspar) is usually pink. You cannot see into a feldspar; it is opaque. But the surface shines (it has a LUSTRE like porcelain), and it has good CLEAVAGE planes in two directions, which means it breaks up into flat-faced blocks.

Feldspars are reasonably hard (6 on MOHS' SCALE OF HARDNESS). On an atomic scale feldspars are made up of ions that fit together into a loose framework. This open framework allows ions to be exchanged easily with liquids that flow past them. This makes feldspars easily WEATHERED, and they are the 'weak link' in many otherwise resistant rocks, breaking down into clay minerals and allowing the rest of the rock to fall apart. Because they are easily weathered, feldspars are found mostly in igneous and metamorphic rocks, and not so commonly in sedimentary rocks (where their weathered product, clay minerals, is far more common).

Micas

Micas break up into flakes. There are many micas, but the most common are brown, white, or colourless muscovite and black biotite. Micas are soft (2–3 on Mohs' Scale of Hardness) but quite resistant to weathering. They are flat, plate-like crystals that normally form small shiny specks in a rock. They are very common in igneous rocks, in many metamorphic rocks, especially schists, and also in some sandstones.

Quartz

Quartz is easy to recognise because it has a glassy appearance. It is colourless when pure, but impurities can give it a variety of colours, especially brown. Quartz almost always forms irregular grains, rather than crystals, with well-defined faces because it has a low melting point temperature and so is the last mineral to crystallise in an igneous rock. It fills all the remaining gaps between the crystals that have already formed. Quartz is quite hard (7 on Mohs' Scale of Hardness), and its compact structure makes it extremely resistant to weathering. As a result, it is found in igneous, metamorphic, and sedimentary rocks. Quartz is the most common mineral in sandstone, where it makes up most of the sand grains.

Ferromagnesian minerals

There is a group of minerals that all contain a large proportion of iron and magnesium, and so they are called the FERROMAGNESIAN MINERALS. They include these groups: olivines, pyroxenes, and amphiboles. In general, they are very difficult to tell apart because they are all green, dark green, or black and often occur only as small crystals. Nevertheless, they form a major part of many igneous and metamorphic rocks, and affect the overall colour of the rock. These minerals weather easily, and so, while common in igneous and metamorphic rocks, they are not common in sedimentary rocks.

Olivine is an olive-green, glassy, hard mineral, crystals of which rarely grow to more than a millimetre across. It is one of the main minerals in basalt and gabbro.

Pyroxene, of which one of the most common minerals is augite, is dark green to black and common in many igneous and metamorphic rocks.

Amphibole, of which one of the most common minerals is hornblende, makes rather long dark-coloured crystals that can sometimes be conspicuous in granite, and can often be more abundant than biotite. Hornblende can be distinguished from biotite by having prismatic (needle-like) crystals rather than plate-like ones.

Garnets are olivine-type minerals – but with a red colour – that form crystals which can often be large, especially in metamorphic rocks, and of dodecahedral (12-faced) shape.

Clay minerals

Clay minerals are secondary minerals, meaning that they do not occur in either igneous or metamorphic rocks, but only as a result of the weathering of minerals in these rocks. Clays are a major constituent of mudstone and shale. Crystals of clay minerals are so small that they cannot be seen, except with the most powerful electron microscope. They are plate-like in structure. Weathered feldspar is the main source of clay minerals.

Calcite

Calcite is a mineral produced by the precipitation of calcium bicarbonate in water. Calcium is the main element in limestone and also in many animal shells. It is extracted by living things directly from seawater and then precipitated as the bicarbonate.

Calcite is colourless or white, but often occurs as grains, rather than as crystals, mixed with other materials, such as clay. In this case the rock looks grey.

Calcite is soft (3 on Mohs' Scale of Hardness), and commonly found in sedimentary rocks. It is also found in metamorphic rocks as marble. Calcite is readily dissolved by water.

Dolomite

This mineral is similar to calcite but contains magnesium as well as calcium, and is more resistant to weathering. It is opaque, dull white and is slightly harder than calcite. It is found in sedimentary rocks.

Evaporites

A number of minerals are formed by the evaporation of seawater. They are collectively called evaporites, of which the most common are halite (salt, sodium chloride) and gypsum (calcium sulphate). They are very soft minerals (2 on Mohs' Scale of Hardness) and occur as thick beds. They are very soluble minerals, only found in sedimentary rocks.

Chapter 3: Igneous rocks

Igneous rocks (from the Latin word *ignis*, meaning 'from the fire') are all those materials that began as liquid rock. They are the most uniform of all the rocks we find at the earth's surface because, as with water being boiled in a pan, molten rock is continually churning and mixing its ingredients. Also, once a rock has become molten, it retains no history of what happened to it in its past.

Igneous rocks typically have their origins at between 50 and 200 kilometres below the crust, although the immediate source of a volcano may, for example, be a magma chamber just a kilometre or so below the surface.

In total, igneous rocks are the most common on the earth's surface. They make up the floor of all of the world's oceans and cover 70% of the surface of the planet. They, therefore, have top place in the rock league table. On land they are mainly found underlying all continents and mountain ranges.

(Below) Igneous materials that flow out onto the surface are called volcanics. This picture shows a basaltic magma forming a small cone. The liquid magma is red, but the cooled minerals are black. The dark colour of the basalt is due to the amount of dark ferromagnesian minerals it contains.

(Below) This is basalt after it has cooled and been exposed to the weather. The colour has changed from black to brown because of chemical weathering of the iron-rich minerals (a form of rusting). The tall columns are caused by the rock shrinking as it cooled. This is Hell's Canyon, Idaho, USA.

Volcanic (extrusive) rocks

Igneous rocks are grouped either as EXTRUSIVE, which means that they reached the surface before they cooled, or INTRUSIVE, which means that they cooled while still inside the crust. Surface (extrusive) igneous rocks include LAVA and ASH.

Molten rock, or magma, may erupt freely out of a volcano and form a tongue of lava that flows for a while, then congeals and turns into solid rock. The magma will be a mixture of a few solid pieces in a large amount of liquid. The liquid will be a mixture of materials, some of which are gases at atmospheric pressure. Thus, as the magma nears the surface, the liquid will start to generate bubbles internally. In freely flowing lava much of the gas in the bubbles will escape, but some small bubbles will remain. They are called VESICLES.

The way a lava flows depends on the amount of silica it contains. If it contains a high proportion of

(Below) This is an example of a lava (andesite) with many gas bubbles.

(Below) The vesicles, or gas bubbles, in this lava were filled with a mineral after the lava cooled. The filled bubbles are called amygdules.

silica, the lava will be sticky and only flow a short distance. If the lava contains a lower proportion of silica, it will flow more freely and form sheets that can cover large areas of ground. A common form of runny lava is BASALT.

(*Above*) Pahoehoe lava being formed on Hawaii.

The surface of a lava may look different from the body of the lava, for it is here that the lava cools most quickly and also where gas bubbles can most readily escape. As a result, the surface of lava is often full of holes, so that it often looks like a sponge that has been turned into rock.

Pumice is the frothy, grey material that forms on the top of some acid magma as it erupts. It is particularly common on top of more sticky lava. This rock is filled with large vesicles and is so light that it will float on water. A term for lava with a rough cindery surface and large empty bubbles in it is SCORIA.

Below the surface, however, the rock will have much smaller bubbles, and rock from a thick lava flow may have no bubbles at all.

(*Below*) Notice the shiny surface on this piece of lava. It indicates a glassy texture.

Crystal formation in extrusive rocks

Magma begins as liquid material with a temperature greater than 1000°C. The speed at which it cools determines the nature and size of the crystals that will form. The formation of crystals is called CRYSTALLISATION. If the rock cools quickly, the crystals will be stopped from growing and will remain small. If the rock cools slowly, the crystals will continue to grow and become large. Thus, by looking at the texture of the rock, we can tell much about its history.

If the magma cools very quickly, there will be no time for crystals to form, and so the material will consist of VOLCANIC GLASS. If the magma cools slightly less quickly, although the majority will still be volcanic glass, a few crystals will have time to grow. This will produce a rock called a PORPHYRY.

Pyroclastic rocks

Gases forming in sticky magma will be able to escape less readily than in runny lava, and they may be able to escape only when they get close to the surface. As a result, the release of gas may lead to an explosive eruption that throws a column of tiny fragments of molten rock high into the air.

As the magma fragments cool, they turn into a dust-like material called ash. This material is loose when it first lands, but over time it consolidates into a rock called TUFF. Ash is known as a PYROCLASTIC MATERIAL, meaning it is made of solid materials ejected from a volcano. Some pyroclastic materials are much larger than ash, and are called cinders if they are pebble sized and volcanic bombs if they are larger. Combinations of materials thrown out of a volcano can become consolidated into a rock. This kind of rock will have angular fragments and will consist entirely of volcanic materials. It is called an AGGLOMERATE.

(Above) An andesite lava with scattered large crystals forming it.

(Above) Explosive eruptions produce agglomerates.

(Below) An agglomerate with many angular fragments of different sizes fused together.

Intrusive rocks

Intrusive rocks cool more slowly than volcanic rocks because of the insulating quality of the rocks which surround them. These rocks also remain under pressure as they form, so no gases are liberated, and there are no vesicles.

Crystals can only grow in magma that is cooling very slowly. This can only happen deep within the earth. Some intrusive rocks are formed close to the surface and make sheets of rock between others, called DYKES and SILLS. Under these conditions the cooling will be relatively quick and the crystals smaller than in cases where large bodies of rock cool more slowly. These rocks may not be fully crystalline, and some may be glassy.

During the slow cooling of rock, the minerals that form depend in part on the temperature of the melt. Some minerals crystallise out of the melt at higher temperatures than others. The crystals that grow first tend to be the best formed, with the most perfect geometrical shapes. Feldspar is usually the first mineral to form crystals. However, most minerals crystallise at between 700 and 600°C, and so another factor must also be involved in stopping the later crystals from growing large. This factor is usually the amount of water left in the magma.

Magma with a high water content is very runny. If the water

(Above) A volcanic ash.

(Below) This is a plug from an ancient volcano. The plug is made of solidified lava, while the cone that formed around it was made of ash that probably formed tuff. The tuff is much less resistant to erosion than the igneous plug, and so the tuff has been eroded away, leaving the plug standing up in the landscape. This is the Le Puy region of France, and the plug has been used as a base for a church.

content drops by a factor of just four, for example, by boiling off, the runnyness decreases by a thousand times. A magma that is sticky will form crystals less easily than one that is runny. So, it seems that a change in the crystal growth is more closely connected with the water boiling off than with the lava cooling.

The crystals that grow later must fit into the spaces within those already formed. Thus, those that form at later and later stages are bound to be less well shaped. The last to crystallise is usually silica (quartz), and this is why it is often seen as a glassy **GROUND MASS** with no definite crystal shapes.

(Above and below) This basalt is a sill exposed at the coast. Because it cooled while in contact with cold rocks, it cooled quickly and has no visible crystals and no vesicles.

Material containing no well-formed crystals is said to be **CRYSTALLINE**. A material that has some very large crystals set in a ground mass of much smaller crystals is called a porphyry. The individual large crystals are called **PHENOCRYSTS**.

(Left) A porphyry made up of orthoclase feldspar phenocrysts in a **MATRIX** of smaller feldspars, black hornblende, and glassy quartz.

Common igneous rocks

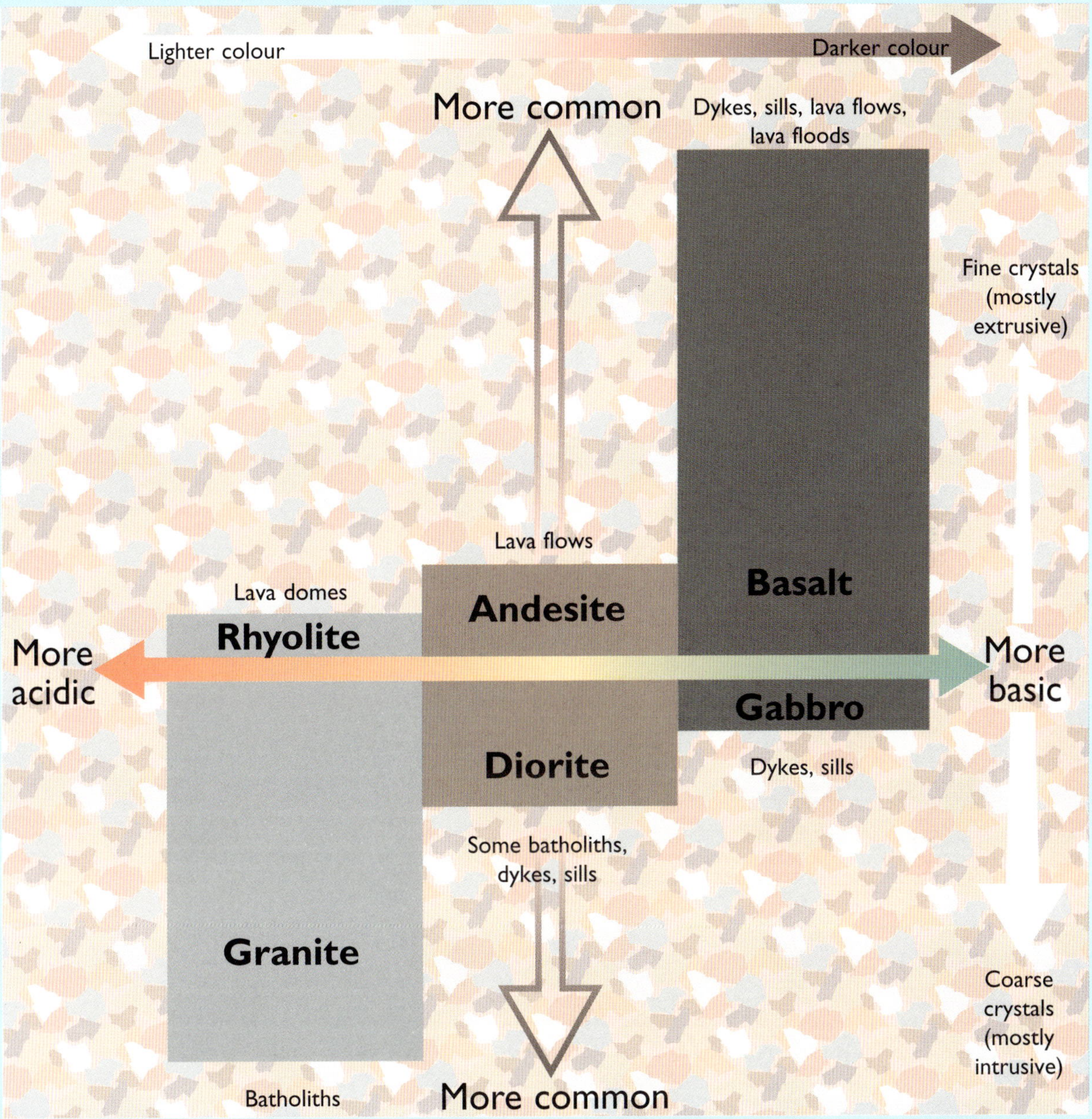

This diagram shows how the various types of igneous rock are related. The upper row shows rocks with crystals too fine to be seen by the naked eye. These are mainly volcanic (but include some dykes and sills where the rock also cooled quickly). The lower row shows rocks with crystals that can easily be seen by the naked eye. These rocks formed slowly at depth.

The columns show the amount of silica in the rocks and the dominant rock colour. The rocks on the left are acid, are composed mainly of feldspars and free quartz, and are light coloured. The rocks on the right have no free quartz, they are dark coloured and contain (ferromagnesian) iron and magnesium minerals, which are all dark.

The length of the bars gives an indication of how common each type of rock is. From this diagram, therefore, you see that the most common intrusive rock with large crystals is a light-coloured acidic rock called granite. The most common volcanic rock with fine crystals is a dark-coloured rock called basalt.

Common volcanic rocks

Obsidian

Glassy rocks are often associated with rapid cooling of very sticky lava. Obsidian has a low water content and a high silica content. It normally forms on the surface of a rhyolite type of magma. This kind of magma, although also very acidic and with a high silica content, contains about 10% water, which keeps it mobile. When the magma reaches the surface, the water boils off. As a result, the magma increases rapidly in stickiness and, at the same time, cools. These are the conditions that produce volcanic glass. Obsidian is, therefore, the material that normally forms on top of rhyolite lava flows. You cannot see through the glass because it contains too many embryo crystals. As a result, there is no clear path for light to pass through.

Rhyolite

Rhyolite is an acidic lava whose mineral content is similar to granite. It is a very sticky lava, associated with explosive eruptions. It does not form extensive flows and quickly solidifies, often before extensive crystals can form. It has a number of glassy variants that form on the surface, including obsidian and pumice. Rhyolite may show flow banding and may even have some visible crystals in it.

Pumice is often formed at the surface of rhyolitic lava as gases try to

Obsidian

Texture: Glassy
Structure: Conchoidal fracture layered bands of colour
Colour: Very dark green, black
Minerals: None, because it is glassy
Origin: Highly acid lava

(Below) A piece of obsidian.

(Left) Rhyolite, a sticky lava.

Rhyolite

Texture: A glassy to finely crystalline rock
Structure: Commonly shows banding where the lava has flowed
Colour: Light grey, pink, and red
Minerals: Quartz, alkali feldspars, biotite, and hornblende. Because it is a lava and cools quickly, there is no time for large crystals to form
Origin: An acidic volcanic rock which flows out as sticky lava. Can also be found as dykes and sills

escape, but become trapped by the VISCOSITY of the lava. The large number of vesicles in pumice gives it the appearance of a brittle, grey-coloured sponge.

(Above) Pumice, showing the vesicles that make it less dense than water.

Andesite

An intermediate type of lava, half-way between rhyolite and basalt, andesite is the most common form of lava associated with explosive volcanoes. Its name comes from its association with the active Andean volcanoes.

Andesite makes up the main material of a composite volcanic cone. Andesitic lava contains many vesicles. It may be light coloured, but is normally dark, especially brown. Although it is not a particularly acid lava, it still has a reasonably high viscosity, and so forms short lava flows that usually develop into tongues that move short distances down the sides of composite volcanoes.

Andesite

Texture: Fine-grained crystalline rock with larger crystals (phenocrysts) of plagioclase and dark-coloured minerals

Structure: Normally no banding

Colour: Grey to brown and pink

Minerals: Very small plagioclase feldspar, hornblende augite, and biotite crystals

Origin: As lava flows from vents and fissures in the flank of a volcano

Basalt

Basalt is the most common kind of volcanic rock. It has a low silica content, a dark colour, and is rich in iron and magnesium.

Basalts are very finely crystalline, often with crystals too fine to see. In some cases, larger crystals of olivine, augite, or feldspar can just be seen.

The surfaces of basalts can be frothy, glassy and scoriacious. Cavities also develop in basalts, and they can later become filled with other minerals.

Although they are found mainly close to the fissures of ocean floors, they are also found on land, for example, actively in Iceland and Hawaii. Historically, basalt has been ejected from the mantle to form enormous floods of lava, which have consolidated into vast black sheets. They are

(Below) A pink-coloured andesite. It does not have a significant number of vesicles.

called basalt plateaus or traps. The name trap comes from the world's largest region of flood basalts, the Deccan Traps, India. Large areas of flood basalts also occur in the Columbia-Snake region of the northwestern United States and in the Paraná Basin of South America.

(Above) Very hot basalt lava flows as thin sheets and solidifies to make a rope-like kind of surface. This is called pahoehoe lava.

Basalt

Texture: Very fine-grained crystalline rock

Structure: Commonly contains many gas bubbles (vesicles). Upper surface may be very vesicular, which is called **SCORIA**. The vesicles may be filled with secondary deposits from the liquids in the lava. The infillings of vesicles are called **AMYGDULES**

Colour: Black or very dark brown

Minerals: Rich in ferromagnesian minerals such as augite, olivine and also contains plagioclase feldspar. Quartz and zeolites often fill amygdules

Origin: basalts are the volcanic equivalent of gabbro and are basic lavas with a low silica content. They are the most common volcanic rocks in the world and can be found in large sheets as plateau basalts and filling dykes and sills. Their surface is either smooth (**PAHOEHOE LAVA**) or rough and cindery (**AA LAVA**)

(Above) Basalt lava that is not particularly hot solidifies on the surface into a scoriacious material that is broken up by the lava flowing beneath. This broken, rough material is called aa lava.

(Right) Unweathered basalt is dark. It can have many vesicles.

(Right) As volcanic rock cools, it shrinks in much the same way as mud cracks as it dries out on the floor of a lake bed. In especially favourable conditions the rocks shrink uniformly around many centres, creating a pattern of cracks that gives rise to columns (see also page 15).

Tuff

Volcanic activity produces a large amount of ash that settles over the landscape. It is particularly thick close to the vent of a volcano. After an eruption has ended, the ash begins to consolidate, and a later covering with lava or more ash will turn the ash into a soft rock called tuff. In effect, the rock could be thought of as a sediment because it is deposited, but the material is entirely igneous, and it forms only at the time of an eruption.

Tuff

Texture: Dense, fine-grained agglomerate made of tiny particles of pyroclastic material. It is usually not very compacted

Structure: None, occurs in thick beds

Colour: Very light grey, pink and white

Minerals: None

Origin: Volcanic ash that has settled during an explosive eruption. It often contains many layers, each indicating an eruption. It is very porous, and percolating solutions can penetrate through it, cementing the grains together to make a welded tuff, which is a hard rock

(Left) A volcanic tuff.

Intrusive igneous rocks

Volcanic rocks are often only a small part of the total amount of igneous rock formed when plates collide. Huge volumes of igneous rock remain buried deep underground, while some magma forces its way in between or across pre-exisiting rock layers, forming dykes and sills. When magma cools underground it forms intrusive rock. (For more information on volcanic activity see the book *Earthquakes and Volcanoes* in this set.)

Plutonic rocks

Intrusive rocks may cool quickly if they are thin sheets of magma squeezed between cold rocks. Dolerite is an example of this kind of fine-grained rock. But large masses of magma – for example the magma plumes that are the sources of volcanoes – cool very slowly, giving time for many varieties of crystal to grow at the same time.

All igneous rocks formed deep underground are called PLUTONIC ROCKS (after Pluto, the Greek god of the underworld). Granites and gabbros are examples of plutonic rocks. Usually, crystals in plutonic rocks are large enough to be seen with the naked eye. The texture is, therefore, coarse, but the grains are of similar size. Phenocrysts are uncommon in plutonic rocks.

Diorite

Texture: Medium- to coarse-grained rock

Structure: Massive

Colour: Like granite, but darker because it is richer in hornblende and augite

Minerals: Plagioclase feldspar, orthoclase, hornblende, biotite and augite with very small amounts of free quartz

Origin: A common rock in dykes leading away from large granite **BATHOLITHS**. Diorite is often the rock which, when liquid, supplied the fluids for **HYDROTHERMAL VEINS**

(Right) This is a piece of diorite. The brown line around part of the edge shows the effect of the weather in oxidising iron-rich minerals. It is, in effect, a layer of rust. To avoid being confused by the effects of weathering, geologists always break open samples to obtain a clean, unweathered surface.

Diorite

Diorite is the typical rock of dykes and sills. This is the intrusive equivalent of andesite.

Diorite is a medium to coarse-grained intrusive igneous rock that contains about two-thirds plagioclase feldspar and one-third dark-coloured minerals, such as hornblende or biotite. Many diorites are the result of a reaction between magma and the surrounding rocks, and often include fragments of foreign rock.

Gabbro

In places where tectonic plates are splitting apart, magma wells up and cools in thick, vertical sheets. This plutonic rock is called gabbro and it has the same composition as basalt. Although gabbro makes up a thick layer below the basalt that covers the ocean floor, it is rare to find gabbro on land. Outcrops of gabbro mainly occur where parts of ancient ocean plates have been buckled upward and made into mountains.

Gabbros are very variable in texture, but generally are medium or coarse-grained rocks that contain mainly plagioclase feldspar and augite or olivine. Rocks called dolerite or diabase arc varieties of gabbro with small crystals.

Gabbro

Sometimes called black granite.

Texture: Coarse-grained rock

Structure: Massive

Colour: Dark grey or greenish black. Can appear mottled

Minerals: Plagioclase feldspar, augite, hornblende and olivine

Origin: Plutonic basic igneous rocks. Often in the form of sills and dykes and other intrusive features. Forms much of the ocean floor

(Below) Dolerite.

(Below) Gabbro.

Granites

The great plutonic igneous bodies that form the core of most mountains are made of granite, or of a slightly less acid and slightly darker variation of granite called granodiorite. They are the most common intrusive igneous rocks on land. Occasionally, granite may form dykes or sills, but all the great intrusive masses that were once magma chambers supplying active volcanoes are granite. These great rock bodies underlie most mountains and are called batholiths.

The granites are coarse or medium-grained intrusive igneous rocks that are rich in quartz and feldspar. The dominant mineral is feldspar (both kinds). If the rock has less than one-fifth of free quartz or more than one-fifth of dark-coloured minerals, it is probably granodiorite rather than granite. The most common minerals in granite, apart from feldspar and quartz, are small crystals of muscovite and biotite micas, as well as small crystals of hornblende or augite.

Granite

Texture: Medium- to coarse-grained plutonic igneous rock

Structure: Usually massive. Occasionally hornblende will separate out in rough bands that make it look somewhat like a gneiss. It forms large blocks as a result of the rock expanding as the original overburden is stripped away by erosion. Sometimes forms domes

Colour: Varies with feldspars but is usually white or light grey or pink (the colour being given by the feldspars), with ferromagnesian minerals producing a speckled appearance

Minerals: Quartz, orthoclase feldspar, plagioclase feldspar, muscovite, biotite and hornblende

Origin: Formed by the slow cooling of magma in magma chambers

Uses: It is a very hard rock and so is widely used for roadstone. It will also take a polish and so is used as a decorative stone. It has excellent weather-resisting properties

(Below) Granite with pink plagioclase and white orthoclase feldspar, black hornblende, and grey glassy quartz.

(Left) Granite with phenocrysts of orthoclase feldspar (white).

(Below) Hornblende granite.

(Below) A granite with large phenocrysts of plagioclase feldspar. This granite is weathering and falling apart as the smaller feldspars change chemically into clay minerals.

Granodiorite

Texture: Medium- to coarse-grained rock

Structure: As granite

Colour: Like granite, but grey or white only

Minerals: Quartz, orthoclase feldspar hornblende, biotite, augite

Origin: A cooled magma like granite. It is a very common rock and only a little less common than granite in batholiths

(Right) Famous Half Dome mountain in Yosemite National Park, California, USA, is composed of granodiorite.

(Below) Granodiorite.

Pegmatite

Some forms of igneous rocks are called pegmatites. They are rocks in which large crystals form. The rocks have the same chemical composition as magmas, but have a high content of water and other volatiles. This makes the magma more runny and it makes it easier for ions to migrate to where crystals are forming, and so for crystals to grow very large. Pegmatites are normally found as dykes, sills, and other igneous intrusions (i.e. offshoots from a magma chamber or vent). Crystals in pegmatites may be over a metre across (as opposed to normal crystals, which may be 2–3 cm across). Pegmatites are often mined for their metal content, for example, copper, gold, silver, and tin.

(Below) Pegmatites typically contain very large, well-formed crystals.

Granite pegmatite

Texture: A granite with very coarse-grained crystals

Structure: Some dykes have cavities in them in which crystals form

Colour: Light grey to white or pink. The crystals are very irregularly organised, which makes them look very different from granites

Minerals: Quartz, potash, and plagioclase feldspar, muscovite, biotite, hornblende

Origin: Commonly found in dykes of granite. The dykes may cut across granite, showing that they were intruded after the main body of granite had cooled

Uses: Pegmatites contain large crystals and so are good for collecting specimens. They are sought after by gemstone collectors of beryl and topaz, among other minerals

Chapter 4: Sedimentary rocks

A sedimentary rock contains layers, or STRATA, formed by fragments of other rocks (sediments) carried by rivers, ocean waters, or wind and laid down in BEDS.

Sedimentary rocks are formed of many strata. The materials in the strata, and the patterns that they form, each help tell the story of how they were laid down. However, a sedimentary rock is like a diary that somebody keeps only now and then. Sometimes there is a long and complete entry; then there are long pauses between entries. And sometimes the 'diary' is tilted on end or even upside-down.

Groups of sedimentary rocks

Sedimentary rocks are divided into three groups: CLASTIC ROCKS (whose name comes from the Greek word *klastos*, meaning 'broken'), which are made of small pieces of rock, weathered from other rocks; CHEMICAL ROCKS, which are laid down by some form of chemical action, such as salt; and ORGANIC ROCKS,

(Below) This diagram shows some of the places where sedimentary rocks might be expected to form.

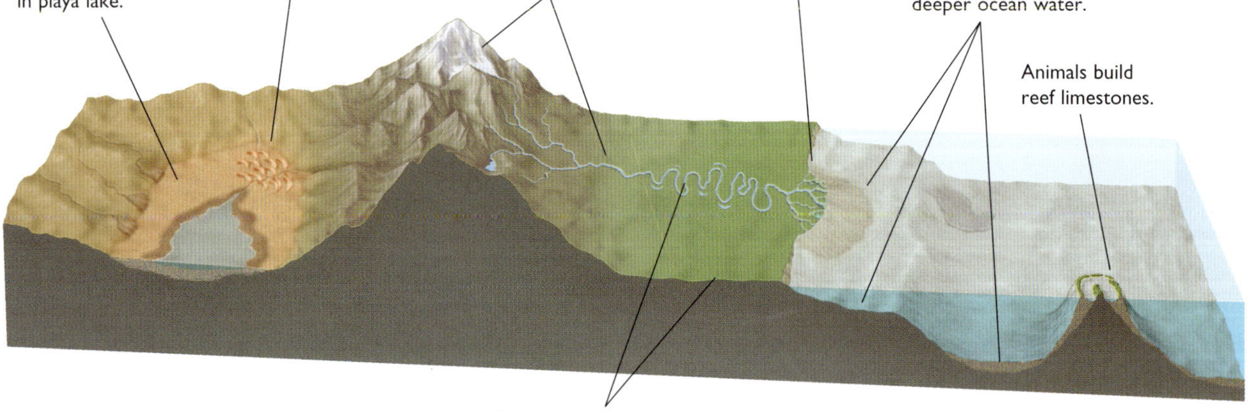

which are formed from materials produced by living things, such as plants (coal) and animals (the shells that make reef limestones).

The difference between clastic and nonclastic rocks is often easy to see because clastic rocks are usually made of many thin strata, whereas the other rocks tend to be single thick beds.

Where sedimentary rocks form

Many sedimentary rocks form in shallow seas close to the edges of continents. Here there are plentiful supplies of sediment brought by rivers carrying material eroded from the land. In some places the coarsest materials form huge deltas. The Mississippi, for example, currently produces about four cubic kilometres (a cubic mile) of new sediment every 20 years. Waves and currents then act on the sediment to rework it, the coarsest material moving along the shore, the finer material often being carried farther out to sea.

Because many sedimentary rocks are made of material carried by rivers, the pattern in which they are laid down varies with the changing pattern of floods and drought on land. A period of flood, for example,

(Below) This cliff is composed of many strata, or rock layers. Each stratum was laid down in a different environment: shales where the water was still; sand where currents were stronger; and limestone when there was no sediment entering the sea at all. The fact that the strata repeat many times shows that over the millions of years that the rocks of this cliff were laid down, conditions changed time and time again.

may flush a large amount of coarse material to the sea, whereas a long dry period may result in only fine material being carried seaward. As a result, sedimentary rocks often contain many layers of varying texture and thicknesses. Sometimes strata may be of coarse material and tens of metres thick; at other times they may be of fine material and just a few centimetres thick.

A certain amount of sediment can build up on land, and locally this, too, can produce rocks. Deserts, river valleys, the foot of large mountain ranges, and near glaciers, are all places where rocks can build up on land.

(Above) Debris on a mountain front can be cemented into a breccia.

Most sedimentary rocks were formed under the sea. Because we find rocks now forming cliffs and even mountains thousands of metres above sea level, this, in itself, tells of the enormous changes that must have happened to the land and sea during geological time.

It is quite common for clastic and nonclastic rocks to be formed in the same place. Limestones, for example, are often interbedded with sandstones (made from beach sands) and shales (rocks made from mud).

(Below) Flint is made from silica precipitated from solutions percolating through the chalk as it formed.

Some sedimentary rocks are very pure – for example, the gleaming white chalk cliffs of Dover, southern England. However, it is more common for limestone rocks to be contaminated with mud. Provided the material still contains more than half its volume as shells and other forms of calcium carbonate, it will still be a limestone. However, its colour will be affected. Muddy limestones (known as argillaceous limestones) are often dark grey.

What sediments tell us

(Above) This picture of the white chalk cliffs of southern England shows that, under some conditions, almost pure deposits can form. The nodules are pieces of flint.

Sedimentary rocks are uniquely valuable in providing a record of which way up they were formed, the conditions under which they formed, and with the help of fossils embedded in them, when they were laid down. All sedimentary rocks are laid down in flat sheets called BEDS, so that if we find sedimentary rocks that are dipping (sloping), we know that they must have been tilted since they were formed.

As each sheet of sediment is laid down, it may produce a distinctive pattern that is a very useful indicator of how it formed. The most common patterns are cross-bedding (also called current bedding) and graded bedding.

Cross-bedding and ripples

Sandstone beds often show a pattern of sloping lines across them. These CROSS-BEDDING patterns slope from the upper to the lower surface of a bed.

(Below) This is a blue-grey-coloured limestone. The colour has been influenced by the amount of mud mixed in with the calcium carbonate.

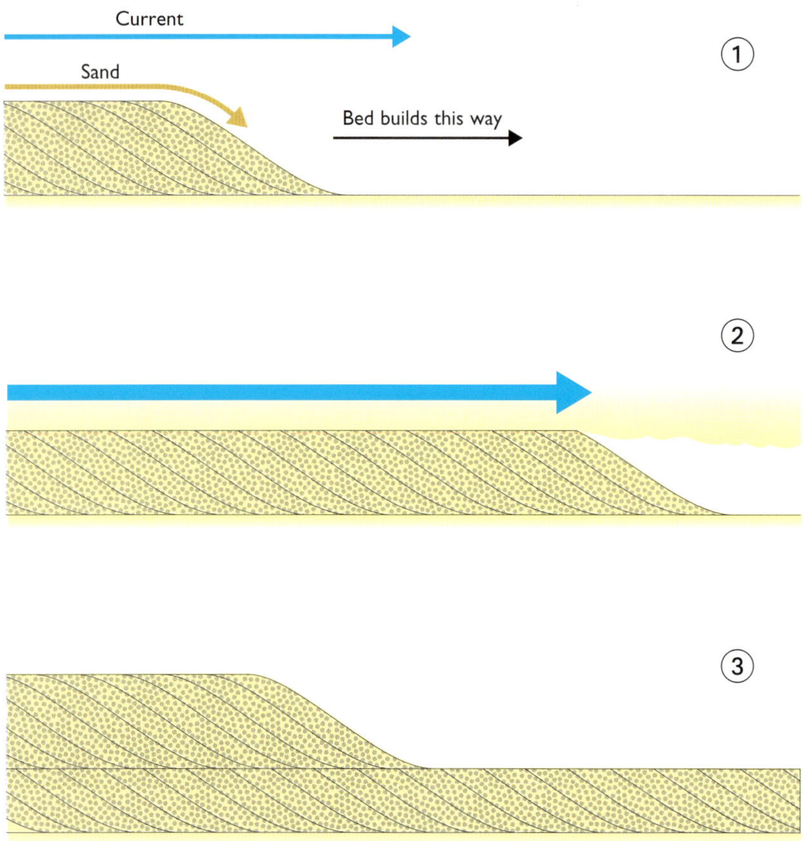

(Left) This sequence of diagrams shows the way in which a new bed is laid down. It might be a river bed or even sand being laid down on a desert.

Notice that the bed builds with the current, the sediment used to build the next part of the bed falling down the steep down-current face.

If a flood (or a severe desert sandstorm) erodes the top of the bed and deposits a new bed on top, we can see this by the change in pattern.

We can also use the pattern to tell which way up the rocks are. This is very useful when looking at rocks that have been folded and possibly overturned during mountain building.

(Below) Zion National Park, Utah, USA, contains some of the most dramatic cross-bedding patterns in the world. These huge patterns were formed by the shifting position of ancient sand dunes.

Rivers, deltas, coastal waves and currents and desert sand dunes can all produce this pattern.

A fast-flowing current often deposits material as it moves. New material is deposited on the down-current side of the old. In this way, new material adds to a bed that builds in the same direction as the current. This is true in a river, on a delta, on a beach, and also for sand building sand dunes in a desert.

Cross beds can be used to find out which way the current was flowing when the material was laid down, and also which way up the beds lay when they were formed.

Ripple marks can also sometimes be seen on the tops of sandstone beds. Ripples are a feature of sand being moved about by shallow waves, as on a beach, or fast-moving currents of water in a river channel.

Graded bedding

As a flood subsides, the heavier and larger particles in suspension will be deposited first, while the smaller and lighter particles remain in suspension for longer. As a result, beds often contain a grading of material, with the coarsest at the bottom and the finest on top.

(Above) Ripple marks on a modern desert.

(Below) Fossil ripple marks on a sandstone.

(Right) This picture shows bedding (horizontal divisions), and jointing (vertical divisions) in horizontally bedded sandstone.

This graded bedding is a way of telling which way up the rocks were deposited. It provides another useful pattern to help identify where beds have been overturned by later earth movements.

Signs of changing conditions

In many cases there is a rapid alternation of beds, perhaps between shale, limestone, and sandstone. This is an indication of changing conditions. Fine muddy sediments can be laid down only when the water is still, whereas sands can be laid down while water is still moving. Limestone can only become an important rock if there are no muds or sands being brought into the area, perhaps because it is an area of shallow sea far from the coast and its sediment-laden rivers, or because the land nearby was a desert. Each change of strata is, thus, an indication that a new phase of sedimentation has begun.

Just because one bed lies on top of another, does not mean that it was laid down immediately afterward. In fact, there may have been very long periods between beds being deposited. Furthermore, the top of the older bed may have been partly eroded before the younger bed was deposited. If a break does occur, it is called a DISCONTINUITY. Such discontinuities cannot be discovered simply from looking at the sediment, however, but only by examining the fossils that the beds contain.

(Below) This is graded bedding, showing the coarsest sediment at the bottom of each bed.

(Below) These are layers of mud that have recently been laid down under water. If compacted, they would form shale rock.

The results of weathering

As rocks break down, all manner of sizes of fragments are released. Some pieces are huge boulders, while others are fine clays. But once released from the bedrock, each of the fragments is acted on by rivers, ice, waves, and wind.

There are two quite different processes that work on rocks: the first is the result of rough handling, for example, as stones are bounced down a river bed or as they are ground along in the bottom of a glacier. It is called MECHANICAL WEATHERING and it can cause sharp edges to be rounded, faces to be polished and large boulders to be worn down to sand grains. In general, the longer weathering takes place, the more the rock is broken down, and the smaller it becomes.

But there is a limit to how small material can become by mechanical weathering. Only one-fifth of all material is of silt size or above – the sizes that can be produced by mechanical weathering. The great majority of sediment – some four-fifths of it – is made of clay, and clay is not a product of mechanical abrasion, but of chemical reactions.

Clay, which forms MUDSTONES and SHALES, is made up of microscopic crystals of clay minerals. They are produced when water reacts with feldspar (the most common mineral in the world). The

(Below) This coarse, poorly sorted material is part of a river bank in a desert. The material is carried by flash floods. If this material were turned into rock, it would form a conglomerate.

feldspar is dissolved, and a new clay mineral produced (chemists would call it a precipitate). Other minerals also weather in this way. This chemical action cannot produce large pieces of material, so mechanical weathering and CHEMICAL WEATHERING are mutually exclusive, with chemical weathering being by far the more important in all but desert and cold conditions.

It is impossible to see chemical weathering at work (except in the rather unusual cconditions of seeing limestone dissolve or be reprecipitated in a cave), but it is easy to see rocks being ground down by water and ice. The importance of chemical weathering is often underplayed. This is also true of the world's rocks, with the greatest amount of rock being shale. Yet because sandstone makes upstanding ridges, while shale produces valleys, the role of shales is also often underplayed.

(Above) A breccia, showing the very large clasts of which it is made. The material is angular and unsorted because it has not been transported far.

Coarse-grained rocks

In general, it can be said that the larger the size of material, the less common it is as a rock-maker. Thus, in the list below, conglomerates and breccias, which are rocks made of boulders, pebbles and gravel, are less common than sandstones, which, in turn, are less common than shales. Siltstones are the least common of all because silt is usually only produced when glaciers grind material during ice ages.

There are two categories of very coarse sediments, called breccias and conglomerates. A conglomerate or a breccia is a rock made of at least one-tenth of coarse fragments of eroded rocks. They are held together by a matrix of finer materials. Breccias are generally made of larger stones than conglomerates.

Breccia

Breccia (which comes from the Italian for fragments of stone) is made of large angular fragments of rock

Breccia

Texture: Uneven, coarse-grained rock with large *angular* pebbles or fragments in a fine cement

Structure: Massive beds with no bedding planes. Can contain fossils, but mainly in the cementing material

Colour: Mainly controlled by the colour of the fragments and of the cement

Minerals: Varies with the fragments and may, for example, be dominated by fragments of gneiss, lava, limestone, or sandstone depending on the local rocks from which it is formed

Origin: It is called a detrital rock because it consists of fragments of other rocks that are unsorted (rubble) that have been cemented together. The rubbly alluvial cones at the foot of a mountain front produce angular material of this kind. Broken pieces of lava in cinder cones can also become cemented together. This is called volcanic breccia.

In general, the gaps between the stones are filled in with smaller material, often of sand size. The term MATRIX is used for this fine material. The matrix and the coarser fragments are then cemented together, often with calcium carbonate that was deposited on the surfaces of thc particles as lime-rich water percolated naturally through the sediment as it formed.

For breccias to keep their sharp edges and corners, they must travel only a short distance. As a result, breccias are found close to the rocks that were used to make them.

Conglomerate

When boulders, pebbles, and gravel travel further, their edges and corners become worn down. Conglomerates are made of anything larger than grit that has completely rounded particles. Conglomerates can form from pebbly beach materials as well as from river beds.

The main agent for movement is running water. The result is that the particles get sorted, with the coarsest traveling a shorter distance than the finest. The presence of all sizes together implies a series of short floods, such as would be common in deserts. The presence of particles of much the same size suggests the sorting may have been on a beach. The cementing agent for conglomerates is normally calcium carbonate, deposited from the waters that seeped through them.

Conglomerate

Texture: Uneven, coarse-grained rock with *round* pebbles in a fine cement

Structure: Massive beds with no bedding planes

Colour: Mainly controlled by the colour of the fragments and of the cement

Minerals: Varies with the fragments, and may, for example, be dominated by fragments of gneiss, lava, limestone, or sandstone depending on the local rocks from which it is formed

Origin: It is called a detrital rock because it consists of fragments of other rocks that are poorly sorted. Conglomerates represent the beds of old desert rivers, old beaches, ridges deposited by ice sheets and other landforms where the material has not been fully sorted

(Below) A conglomerate contains mainly materials above 2mm in diameter and often many stones, but, unlike the breccia, shows signs of sorting and other evidence of being transported by water.

Sandstone

Sandstone, the most common coarse-grained rock, is cemented grains of sand. Another term for sand-sized rocks is **ARENACEOUS** rocks. A somewhat coarser size of material is called **GRIT**, and the rock is called a gritstone. If feldspar grains are present, which might happen, for example, with a sandstone formed in a desert where there is limited chemical weathering, the sandstone is termed **ARKOSE**.

Sandstones make up between a tenth and a fifth of all rocks forming on earth (the proportions change with the world's climates). But they are resistant to erosion and therefore tend to make prominent cliffs and other features in the landscape. They are, therefore, more noticeable than their realtive abundance would suggest.

Sandstones have been formed at all periods of geological time, both in deserts and on coasts. They are dominated by grains between 0.06 and 2 millimetres across.

Most sand, or grit-sized grains are made of quartz (the mineral silica). Quartz is very hard and is among the least reactive of minerals. As a result, it is not

Sandstone

Texture: Coarse-grained clastic rock with even particle sizes. Particles well rounded if they have been on a beach or in a river, and more angular if they have been in a desert. Cemented by silica, carbonates, or oxides

Structure: Well bedded, with layers of variable thickness, often cross-bedded and sometimes with ripple marks. Fossils less common than many other sedimentary rocks

Colour: Colour depends mainly on the cement and can vary from white (with silica and carbonate cement) to dark brown or red (iron oxide cement). Glaucoinite makes the sandstone green

Minerals: Mainly quartz, but feldspar in desert sands (arkose sandstone) and the weathered and rounded fragments of minerals such as garnet may survive

Origin: Detrital origin, and represents delta, desert sand dune, or sea beach environments

(Below) A detail of a sandstone

(Above) This butte is made of many beds of sandstone, each produced in different conditions and showing different textures and colours. The differences are highlighted by the variations in resistance to erosion.

readily broken down either by abrasion or by chemical means.

As quartz grains roll about in rivers or are tossed about by coastal currents, their edges are smoothed down and become more rounded in shape. However, they do not change chemically and so remain to form the skeleton of many later rocks. This is why sandstones are among the most common rocks.

The sand grains in sandstone are usually cemented together with calcium carbonate, silica, and iron oxide. The durability of sandstone is, therefore, not usually related to the hardness of the quartz, but to the cement. If the cement is thin, sandstones can be weak and crumbly. If the sandstones are subject to a large amount of percolating water, the calcium carbonate can be dissolved away. The hardest sandstone rocks are cemented with silica (in which case they are usually called quartzite) or with iron oxide (in which case they usually have a strong brown or red colour). Glauconite cement turns the sand green and the rock is called greensand. If the cementing agent is clay the rock is called ARGILLACEOUS. The amount of pore spaces filled up with muds and cements determines how porous the sandstone is.

(Above and below) This is a gritstone, a very coarse form of sandstone. It shows current bedding.

(Below) A sandstone showing layers. The cementing agent is iron oxide, producing an orange colour.

Sandstones form in two quite different environments: in water and on land. On land they are made from wind-blown sand, as in a desert. Sandstones deposited under water were formed either as a beach deposit or as part of a delta. These sandstones occur as thin layers, although many also show ripple marks to indicate they were once a beach.

Cross-bedding is another common pattern found in sandstones. Some cross-beds from sand dunes are tens of metres thick, but cross-beds from river and sea environments are quite thin.

Sandstones can also show graded beds. Here the coarsest material is at the bottom of the bed with the particles getting finer higher up. The bottom of each bed is marked by coarse bedding. This may indicate changes in the environment that have occurred over time, with swift currents being replaced by slacker currents before a return to swift currents again. Sandstones formed in water also often contain fossil remains.

Sandstones often form in rapidly changing environments, so that layers of sand may be interbedded with layers of silt, for example. In such cases the harder sandstone makes sheets that split apart along the siltstone layers. They are known as flagstones.

Greywackes

A special term, wacke, or greywacke, is used for dark-coloured, very strongly bonded sandstones that contain both larger particles and a large proportion of mud matrix. They contain no sorting and no bedding. Almost all wackes were deposited in deep water by the collapse of sediments, probably on the edge of CONTINENTAL SHELVES.

Wackes are found only in rocks close to mountain systems and ocean trenches. This is because they indicate a period of very rapid and unstable deposition – for example, close to a rapidly eroding mountain chain.

(Below) A sandstone cemented by calcium carbonate. This sandstone is much paler than one cemented with iron oxide.

(Below) An arkose sandstone.

(Below) A wacke. It is a poorly sorted rock and may have both sand and clay particles.

Fine-grained rocks

Fine-grained rocks are the world's most common continental rocks. Shales and mudstones are the most common sedimentary rocks in the world, making up four-fifths of the sedimentary rock.

(Above) Shales are uniform fine-grained rocks that fracture into sheets, which are more or less flat, but have curved surfaces. They are all very soft.

Mudstone and shale

Mudstones and shales are made of the finest sizes of material, either silt or clay, both of which are too fine to be seen with the naked eye. As a result, mudstones and shales appear to be very uniform dark-coloured rocks. Darker colours indicate a larger proportion of carbon material. If they incorporate a lot of organic matter they can become oil shales and coal shale. Calcareous shale has a large amount of calcium carbonate in it; if the proportion of calcium carbonate increases to over one half, the rock will be called a limestone.

Bedding planes may be more or less distinct, with the material breaking up into flattish sheets.

Essentially, the clay and silts that make up mudstones and shales come from the erosion of other types of rock. However, about a third of the rock is usually made up of other materials, for example, calcium carbonate (about 5%) or silica (up to 30% as cement between the grains) and also organic matter (about 1%) derived from buried plants and animals. Iron compounds may also make up to 1%. All shales contain good examples of fossils because the quiet waters in which they were deposited are very suitable for the preservation of the bodies of living things.

Shales form in deep ocean basins as well as in the outer parts of continental shelves, and in some sheltered bays, deltas, and lakes. Clays entering the sea from rivers often clump together (flocculate) as they meet seawater, and this allows them to behave like larger particles and settle closer to the coast than would otherwise be imagined. The widespread existence of mudflats in estuaries shows this effect.

Shales and mudstones are made from the smallest

Shale and mudstone

Texture: A fine-grained rock with dense texture having only silt or clay-sized particles. In general, they are too small to be seen with the naked eye. Mudstone is less well compacted than shale. Mudstone absorbs water and breaks up when wet. Both mudstones and shales are very soft rocks

Structure: Mudstone is massive, while shales are more thinly bedded. Shale breaks along curved surfaces into shell-like pieces. Shales and mudstones are sometimes full of fossils

Colour: Grey, the tone of grey depending on the amount of organic matter. May be brown if there is a significant iron oxide content. May be black if there is a significant organic matter content

Minerals: Dominated by clay minerals such as kaolinite

Origin: A detrital rock formed in calm oceanic water, lakes and also in estuaries where the clays stick together and sink. Sometimes layers form, indicating year-by-year changes

(Left and below) Shales commonly fracture into many small pieces. They are physically weak; but since they are made of clay minerals (which are a product of weathering), they are resistant to chemical weathering.

particles and so have the smallest pores. When they are compacted and cemented into rock they are, therefore, often IMPERMEABLE (water resistant).

It is very common to find beds of shale, sandstone, and limestone together. Shale usually makes up the majority of the rock thickness, with the tougher limestone and sandstone making up minor layers. However, because they are more resistant to erosion, these thin beds often act to support the whole rock face, so that it resembles a staircase, with the risers made of tough sandstone and limestone, and the treads made up of weak shales.

Many muds contain the entombed remains of microscopic organisms which, when they decay, may release valuable oils and gases. Black oil shales, for example, are the world's most common oil-bearing rocks. Shales are also the best environments in which the remains of ancient organisms can be preserved. Many shales are very fossil rich.

(Below) Ammonite preserved in shale.

(Left) These cliffs show the typical staircase of bands of weak shale held up by bands of stronger limestone.

When cliffs are formed of shales, they are easily eroded by running water as well as landslides. The landscape they produce in semiarid regions is called badlands. It carries no surface vegetation.

Limestone

Limestone (and the related rock called dolomite) is mainly made up of calcium carbonate, often as crystals of the mineral calcite. Most limestones are made of very small grains 0.001 millimetres across or less. Very often these grains are the shells of microscopic sea animals. Limestone can be formed by the accumulation of vast numbers of tiny shells (for example, lime-secreting algae) or from the erosion and redeposition of limestones.

Limestones contain over half their mass as calcite. They mainly form in shallow seas and are particularly widespread during times when warm, shallow seas cover large areas of land. Modern limestones are still forming in a band close to the equator.

Limestones are made up of three kinds of material. The larger grains that occur throughout the rock are embedded in a fine material made of tiny particles. All of it is cemented together by calcium carbonate. Thus, limestone is similar, for example, to sandstone in the way it is made up of different-sized particles stuck together with cement.

The largest particles of limestone are called ooliths (after the Greek word *oon* for 'egg'). They are

Limestone

Texture: Fine-grained, dense rock with matrix particles so small they cannot be seen with the naked eye. Rarely formed by running water, but mostly as a chemical precipitate. Fossiliferous limestones contain good-sized fragments of fossils, which are often the most conspicuous feature. Oolitic limestones are distinctive for being made of small limestone balls

Structure: May be thin to massive

Colour: Grey to yellow to white. Yellow is the effect of iron oxide staining. Most limestones are grey and some dark grey. Chalk is always white

Minerals: Calcite and aragonite dominate. A limestone must be at least half made of these minerals. Clays form the majority of the remainder

Origin: Chemical precipitate, occasionally detrital, and also formed by organisms in the case of reefs.

about the size of a sand grain and form when small particles roll around in a shallow lime-saturated sea.

The cement is produced mainly by algae which develop calcite needles inside their tissues. When the algae die, the needles fall to the sea bed and soon recrystallise as calcite, locking any limestone grains together.

In some parts of the shallow, warm seas where limestone forms, conditions are suitable for the growth of coral reefs. The reefs make one type of limestone in which the spaces between the corals are filled in with a limey mud. When waves break over the coral, much of it is broken off, so that many limestones are made up of fragments of coral rather than massive reefs.

Another source of limestone is the sea-floor animals called crinoids. They, like the corals, form the starting point of a limestone. The small pieces of a crinoid – known as ossicles – often become buried in

(Above and below) Oolitic limestone is made when grains of sand roll around in a lime-rich sea.

(Above) Detrital limestone formed of many shells and fragments of reef cemented by calcite.

(Below) Shelly clastic limestone.

a limy mud. (For more information on these and other fossils see the book *Fossils* in the *Earth Science* set.)

Most limestones are hard, but one form – chalk – is soft (see photo, page 35). Limestone is normally greyish-white, or even dark grey if it contains a lot of mud and plant remains, but in the case of chalk it is sparkling white.

If hot magma is intruded into or below limestone, the limestone is often altered by mineralising solutions rising up from the magma. Metal ORES, especially zinc and lead, commonly precipitate out in the limestone, while fluorite (calcium fluorite) crystals commonly form in cavities within the limestone.

(Below) Limestones are also common as hard bands of rock among bands of other sedimentary rocks such as shales and sandstones. Here they make the hard, blocky rock bands that stand out from the road cutting.

Coal

Coal is an organic rock which has been altered slightly by being compressed and heated. It, therefore, has some properties that also make it similar to a metamorphic rock. However, since it occurs in beds, it is normally thought of as a sedimentary rock. It was formed by the rapid burial of large numbers of plants. Most coal comes from the burial of swamp forests.

It takes about one and a half metres of plant remains to make each 30-centimetre thickness of coal seam. Many seams are very thin, but some are thick. The Mammoth coal seam in eastern Pennsylvania, USA, is about eleven metres thick.

Plant material decays very rapidly as soon as the plants die. Their preservation in oxygen-poor swamps is, therefore, an unusual event, and strata of coal – called coal seams – are not common. They were mainly created during times when sea levels fluctuated frequently.

The result of repeated changes in the height of the sea was to expose land on which plants could grow, then bury it again, locking the plant remains below a new layer of sediment. This happened time after time during some periods of the geological past. A repeated pattern of this kind is called a **CYCLOTHEM**.

Typically, the beds, including coal seams, show a pattern of land-based material and marine material. Sometimes the coal (which once was a swamp in which trees lived and died) is covered by shale (from a layer of marine mud), sometimes by sandstone (from an ancient delta or beach) and sometimes by limestone (from a shallow sea in which the bodies of sea creatures could build up).

Coal-bearing rocks often show repeating sequences of seams and other rocks stacking up hundreds of metres. Each cyclothem may be ten metres thick.

Coal

Texture: Uniform black rock, often with visible remains of plants

Structure: Occurs in seams. Its hardness varies with how much it has been altered. Anthracite is the hardest and most altered; bituminous coal is less altered and softer

Colour: Black

Minerals: Carbon

Origin: Formed by the metamorphism of plant remains. Chemically the process is called destructive distillation, whereby many volatile materials are lost, and thus the carbon content increases

(Left) A piece of bituminous coal.

(Below) This road cutting shows a repeating sequence of beds quite typical of conditions in which coal was laid down. The dark bands are the thin coal seams.

Evaporites

Evaporites are thick beds that formed as the result of precipitation of minerals dissolved in lagoon or lake waters. They always indicate conditions in which evaporation dominates. The evaporites are mainly rock salt (halite), gypsum and anhydrite.

Evaporites form thick to massive beds, often showing layering where thick halite layers are separated by thin bands of clay representing the sediment from a flash flood. All evaporites are soluble and soft.

(Right) Rock salt, entirely composed of halite.

(Below) Coal is typically laid down in a coastal swamp where the sea level changes regularly and creates a rhythmic sequence of beds called a cyclothem. Many kinds of cyclothem are possible, as shown by comparing this diagram with the sequence in the photograph on the left.

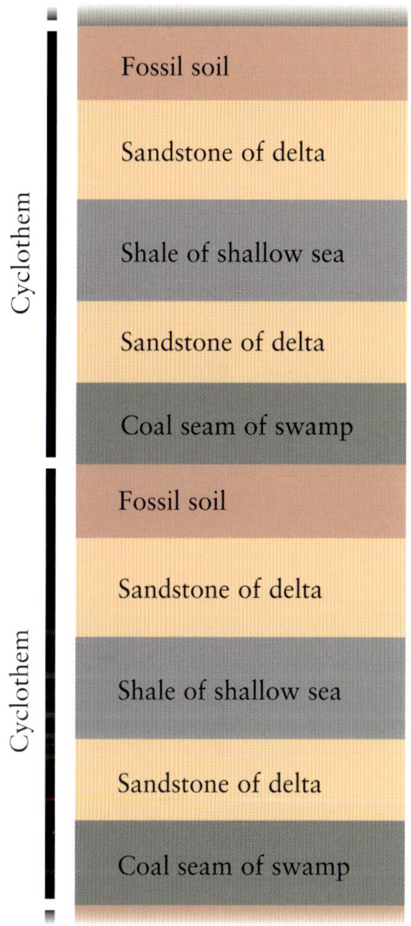

Rock salt

Texture: Crystalline rock with grains visible to the naked eye

Structure: Often forms in massive beds

Colour: White, colourless, pink, and even red depending on impurities

Minerals: Halite

Origin: Lagoons and lake beds saturated in salt water

Chapter 5: Metamorphic rocks

There are many rocks that do not fit into the category of either igneous rocks or sedimentary rocks. Often they have characteristics of both. That is, they often have some kind of layering, but at the same time contain well-marked crystals. They are metamorphic rocks (from the Greek words *meta* meaning 'change', *morph* meaning 'shape').

The origins of metamorphic rocks

Metamorphic rocks were previously either sedimentary or igneous rocks, but they have been subjected to very high pressure and temperature, and to the chemical action of hot fluids passing through their pores. The temperature was not high enough to make them melt entirely and so lose their layered pattern, but it was high enough to allow some crystals to grow, and for the minerals to begin to recrystallise, and thus form a new rock.

The new minerals that are formed under these temperatures and pressures cannot be made in other environments, which helps distinguish metamorphic from igneous rocks (that have crystallised from molten magma). Usually, however, there are more tell-tale signs that the rocks have been greatly disturbed,

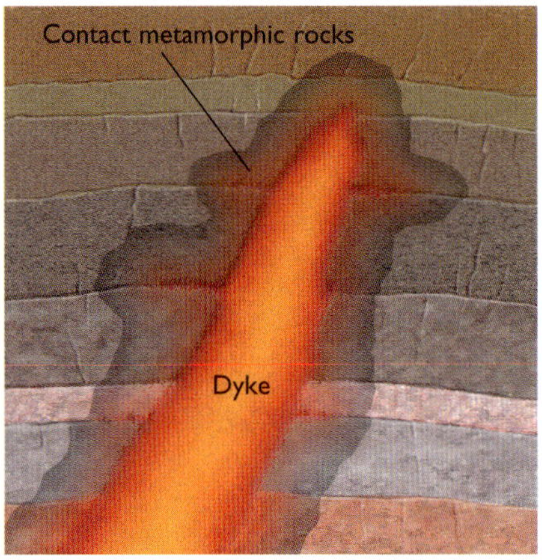

(Below and above) These diagrams show the two types of environment in which metamorphism occurs. The more common is known as regional metamorphism and occurs over large regions where two plates collide and form mountains. The pressure that causes the metamorphism is not simply the weight of overlying rocks, but the active pressure from the plates pushing together.

The lower part of the diagram shows contact metamorphism, or baking, due to the presence of a body of liquid rock – for example, a dyke, a sill, or a magma chamber.

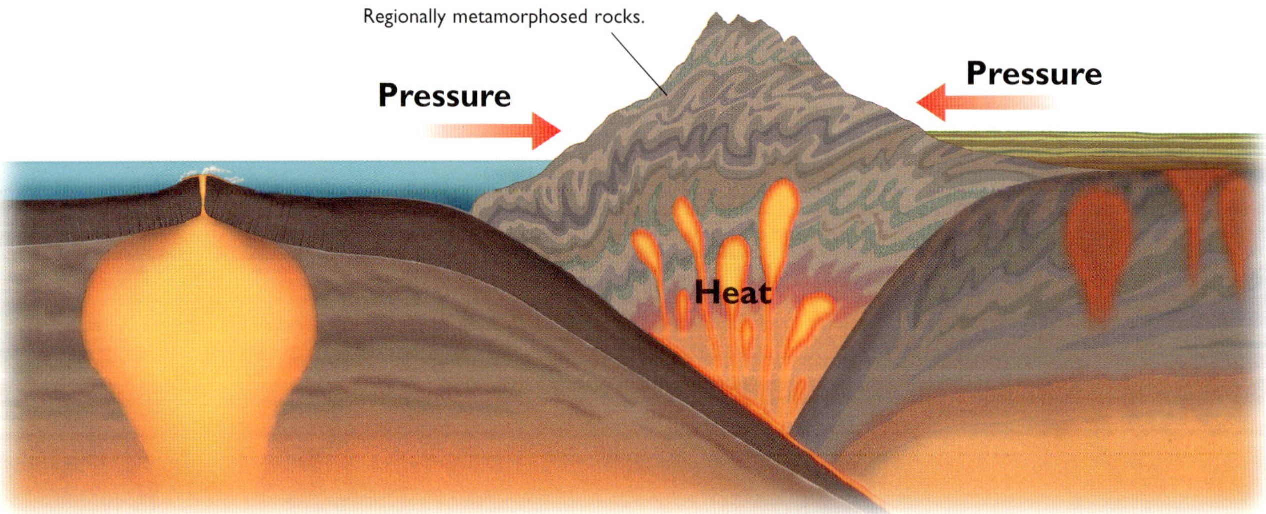

for metamorphic rocks mainly occur in places where there has been a great deal of folding and twisting.

The types of metamorphism

Heat and pressure are applied to rocks in two different ways. In one case the heat comes from contact with molten magma. This is called CONTACT METAMORPHISM, and it is quite local in its effect. The main contact metamorphic rock is hornfels.

The second way is when rocks are forced towards the mantle during the formation of a mountain range. In this case – known as REGIONAL METAMORPHISM – very large volumes of rock are altered.

Metamorphism does not take place on the earth's surface. In the case of regional metamorphic rock, it occurs many kilometres below the surface. We only see metamorphic rocks, therefore, when erosion has stripped away any overlying rocks. So, when we see a

(Above) This is a roadside cutting through part of the core of a mountain belt in the Rocky Mountains, USA. You can see granite (pink) that has melted its way into contorted and intensely folded metamorphic rocks.

large area of metamorphic rocks, we know we are looking at the core of an ancient mountain range.

The results of metamorphism

The changes that take place during metamorphism make the rocks exceptionally hard. Once formed, they are not easily reworked and stand up to erosion very well.

Metamorphic rocks are also important and unique for another reason. The same environments that created them also made it possible for minerals to become concentrated. They are, therefore, the sites of some of the world's most important mining resources.

Crystals in metamorphic rocks

The most common feature of metamorphic rocks is that they are crystalline. They consist entirely of crystals (just as igneous rocks do), but in this case the crystals are not necessarily interlocking and they tend to all face in the same direction, usually at right angles to the direction from which the pressure was being applied.

One of the most striking things about many metamorphic rocks is the growth of individual crystals. Red-coloured crystals of the mineral garnet are especially common in the metamorphic rocks called schists. This tells us that, although the rock never melted fully under pressure, atoms from the minerals in the original rock were able to move around and so grow into new crystals.

To understand how solid rock can flow and grow crystals, compare the way in which rocks form deep in the heart of mountain ranges with red-hot steel forced between the rollers in a steel mill. The steel never melts, but under pressure it readily changes shape.

(Below) How a conglomerate might be changed by metamorphism. Notice the pebbles have been squashed by heating them until they became plastic enough to deform. The fine matrix has re-formed into wavy bands.

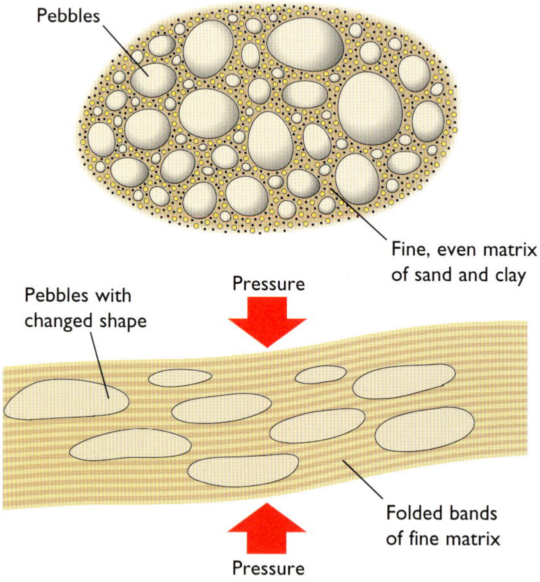

(Below) How a granite might be changed by metamorphism. Notice the crystals have completely disappeared, and instead the rock contains contorted bands of light and dark minerals.

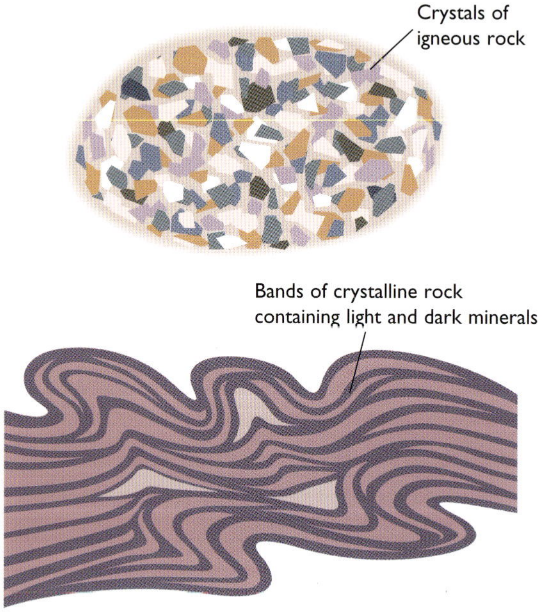

Common metamorphic rocks

There are many kinds of metamorphic rocks because there are several ways in which rocks can be changed: depending on the amount of heat, the amount of pressure, the original texture of the rock being changed, and the minerals in that rock.

As heat and pressure increase, the metamorphic rocks produced occur in this sequence: shale, slate, phyllite, schist, and gneiss. At each stage the texture becomes coarser and the amount of metamorphism greater.

Notice that it is easier to tell that a surface rock, such as sandstone or shale, has been metamorphosed because the changes in them due to heat and pressure are probably greater. Granite, though, which formed from molten rock and, therefore, has already been through events even more drastic than metamorphism, might be almost unaffected. Many gneisses and granites may look similar at first glance.

Regional metamorphic rocks are very common, but are confined to places where mountains once formed. Thus, they underlie the centres of the great continents (which are the roots of worn-down mountains) and also the regions where modern mountain ranges occur. Most plains may be covered with layers of sedimentary rocks, but below them almost certainly lies a basement of metamorphic rocks.

Metamorphic rocks, and granites that lie within them, make up more than four-fifths of all the earth's continental rocks.

Slate and phyllite

Slate and phyllite are the least altered forms of metamorphic rock. They are very fine-grained metamorphic rocks that have been formed from shales and mudstones. The pressure has caused all of the tiny clay minerals to line up in the same direction,

Hornfels

Texture: Hornfels is a fine-grained, even-textured rock that is a baked metamorphic shale.

Structure: Massive and breaks into blocks with curved faces

Colour: Dark grey, brown, or black, often with spotting where minerals have begun to form. Spotting indicates closeness to the igneous rock, causing the baking

Minerals: Microscopic in size and often the same as in the original rock, for example, clays in shale

Origin: Formed in the contact zone of an igneous intrusion – for example a basalt dyke or granite batholith. This zone surrounding a large igneous body is called the **METAMORPHIC AUREOLE**. Hornfels is the most common rock in the aureole.

Hornfels is not the result of widespread pressure as in regionally metamorphic rocks, and the mica grains do not form any kind of banding, and so the rock does not shine in the light, but appears rather dull

(Below) Hornfels, one of the most common contact metamorphic rocks.

and this is what gives the perfection of splitting that makes slates easy to recognise. The splitting in slate is called SLATY CLEAVAGE. Roofing slate is a very good example of this process. The main minerals in slates are tiny muscovite micas and chlorites. As pressure increases, the minerals begin to grow and then separate. This is when phyllites develop; they are half-way to schists.

Slate, phyllite

Texture: Very fine-grained rocks produced when shales were compressed. No crystals visible to the naked eye

Structure: The minerals grew at right angles to the direction of pressure. The result is that the rock tends to break up into sheets. This is called slaty cleavage

Colour: Dark grey. Sometimes green if a large amount of chlorite is present

Minerals: Mica, mostly biotite and chlorite

Origin: Under pressure the clay minerals of the shale re-formed into minute new plate-like minerals such as mica and chlorite.

In slates the original bedding of the shales can often be seen separating one band of slates from another. The original bedding planes and the slaty cleavage are usually quite different. Phyllite is intermediate between slate and schist, and so shows larger crystals, often big enough to be spotted by the naked eye

(Below) Phyllite.

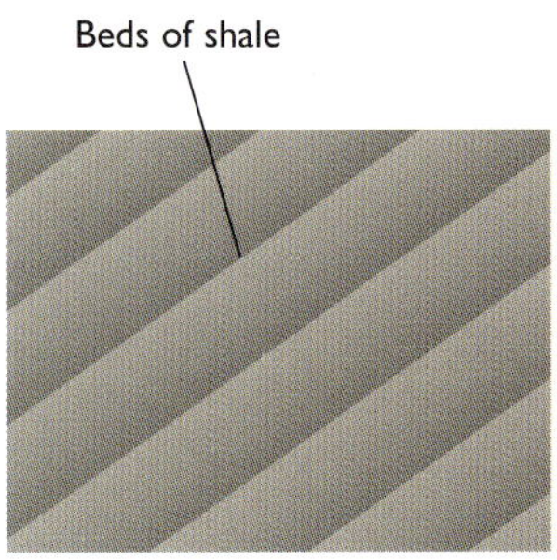

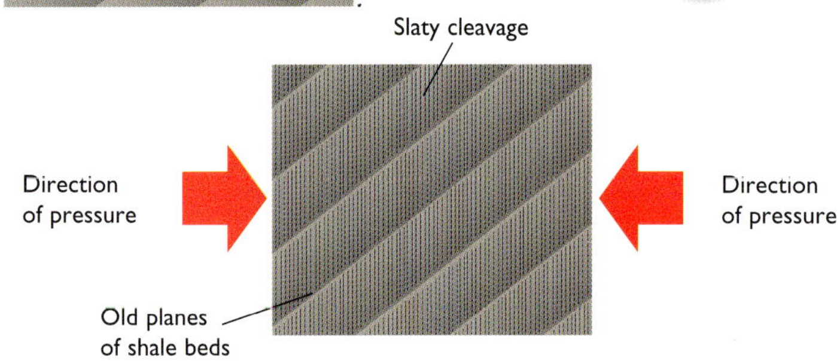

(Below) Slate.

(Below) Marble.

Marble

Marbles are carbonate sedimentary rocks that have been subjected to heat, pressure, and hydrothermal fluids. Because calcite and dolomite minerals are uniform in character, the effect of metamorphism is not to produce a parallel arrangement of the crystals, as happens in slate or schist, but rather to fuse the entire rock together and to increase the grain size. The metamorphic character of the rock is often best shown by the smeared-out colour impurities that give the rock its 'marbling'.

Marble

Texture: Even, granular texture

Structure: Often forms in massive beds. No foliation. Bands and streaks (marbling) occur if impurities are present and show the plastic flow that develops during metamorphism

Colour: White with green-blue and even red marbling depending on impurities

Minerals: Calcite or dolomite

Origin: Marble is made from only one mineral type, and so the rock cannot become foliated as it is metamorphosed. Instead, large grains grow and interlock or fuse into a solid mass. This is what makes it so hard. The marbling lines and patches of dark colour are produced by impurities being concentrated during metamorphism

(Below) Typical shiny surface of mica crystals in a schist (see page 58).

Schist

In many cases, the deep burial of rocks and the vast pressures created, made the layers of rocks begin to creep, or shear, past one another. This causes the rock layers to develop thin sheets called FOLIATION. At the same time, clay minerals begin to change and re-form as medium-sized muscovite mica crystals at right angles to the direction of pressure. This has the effect of making the layering even more prominent. The result is the reflective rock called schist. The shininess is caused by the presence of many muscovite crystals aligned in the same direction.

Schist

Texture: Medium- to coarse-grained rock, showing parallel arrangement of crystals – foliation
Structure: Foliation and reflectivity from the individual mica or chlorite crystals are the best ways to identify a schist
Colour: Silvery, brown if the minerals are oxidised due to exposure (the brown is rust), green if chlorite is present
Minerals: Muscovite, chlorite, garnet
Origin: Schist rock (schist comes from the Greek *schistos*, meaning 'divided', a reference to the way the crystals are arranged in parallel foliation). Schists were produced by more intense heat and pressure than slate, and so the mineral crystals, although often the same as slate, have grown bigger. At the same time, new high-temperature minerals have developed, especially garnet.

Schists may form from shale but can also originate from other types of rocks including basalt, granite and sandstone. As a result, schists are the most common metamorphic rocks.

Crystals do grow large in this type of rock. Garnets are the most common large mineral to form. The garnets in these garnetiferous mica-schists look like eyes, and so this kind of schist is also sometimes called augen schist (from the German *augen*, meaning 'eyes').

The least schistose rocks contain chlorite; then as the amount of heat and pressure increases, garnets begin to form, and finally kyanite crystals grow in the highest-grade metamorphism.

(Below) This piece of foliated schist shows a wavy banding of plate-like crystals of chlorite and mica.

Set inside the foliations are red crystals of garnet. They have obtained the atoms for growth from the surrounding minerals. This rock is known as a garnetiferous schist. The green-coloured mineral is chlorite.

Gneiss

The distinction between schist and gneiss is a matter of the amount of foliation. Schistose rocks are more than half-foliated, while gneisses are intensely metamorphosed rocks that have more than half their structure in a granular form, rather than foliated.

Many gneisses retain the feldspar and quartz from the parent granites from which they form. Others do not. Gneisses may look more glassy than schists, a feature which comes from the amount of quartz in the rock.

(Above) Close-up view of a gneiss, showing granular texture.

(Below) Augen (eye) gneiss. The eye-shaped lens of feldspar is the eye.

(Right) Banded gneiss.

Gneiss

Texture: Gneiss is a coarse, uneven-grained metamorphic rock

Structure: Banding is common as an alteration of dark and light minerals. Bands of quartz and feldspar alternate with bands of mica. The granular layers often contain the remnants of granites from which they have formed. Phenocrysts in the granular layer make the eyes often seen in a form of gneiss called augen gneiss (*augen* meaning 'eyes' in German).

Gneiss is typically very twisted and contorted.

Colour: The light-coloured layers are quartz and feldspar, and mica and ferromagnesian minerals make up the dark bands. Colours are very variable, however

Minerals: Quartz, feldspar, micas

Origin: They result from enormous heat and pressure deep within mountain systems forming during mountain-building periods

Glossary

aa lava: a type of lava with a broken, bouldery surface.

abrasion: the rubbing away (erosion) of a rock by the physical scraping of particles carried by water, wind or ice.

acidic rock: a type of igneous rock that consists predominantly of light-coloured minerals and more than two-thirds silica (e.g. granite).

active volcano: a volcano that has observable signs of activity, for example, periodic plumes of steam.

adit: a horizontal tunnel drilled into rock.

aftershock: an earthquake that follows the main shock. Major earthquakes are followed by a number of aftershocks that decrease in frequency with time.

agglomerate: a rock made from the compacted particles thrown out by a volcano (e.g. tuff).

alkaline rock: a type of igneous rock containing less than half silica and normally dominated by dark-coloured minerals (e.g. gabbro).

amygdule: a vesicle in a volcanic rock filled with secondary minerals such as calcite, quartz or zeolite.

andesite: an igneous volcanic rock. Slightly more acidic than basalt.

anticline: an arching fold of rock layers in which the rocks slope down from the crest. *See also* syncline.

Appalachian Mountain (Orogenic) Belt: an old mountain range that extends for more than 3000 km along the eastern margin of North America from Alabama in the southern United States to Newfoundland, Canada, in the north. There were three Appalachian orogenies: Taconic (about 460 million years ago) in the Ordovician; Acadian (390 to 370 million years ago) in the Devonian; and Alleghenian (300 to 250 million years ago) in the Late Carboniferous to Permian. These mountain belts can be traced as the Caledonian and Hercynian orogenic belts in Europe.

Archean Eon: *see* eon.

arenaceous: a rock composed largely of sand grains.

argillaceous: a rock composed largely of clay.

arkose: a coarse sandstone formed by the disintegration of a granite.

ash, volcanic: fine powdery material thrown out of a volcano.

asthenosphere: the weak part of the upper mantle below the lithosphere, in which slow convection is thought to take place.

augite: a dark green-coloured silicate mineral containing calcium, sodium, iron, aluminium and magnesium.

axis of symmetry: a line or plane around which one part of a crystal is a mirror image of another part.

basalt: basic fine-grained igneous volcanic rock; lava often contains vesicles.

basic rock: an igneous rock (e.g. gabbro) with silica content less than two-thirds and containing a high percentage of dark-coloured minerals.

basin: a large, circular, or oval sunken region on the earth's surface created by downwards folding. A river basin, or watershed, is the area drained by a river and its tributaries.

batholith: a very large body of plutonic rock that was intruded deep into the earth's crust and is now exposed by erosion.

bauxite: a surface material that contains a high percentage of aluminium silicate. The principal ore of aluminium.

bed: a layer of sediment. It may involve many phases of deposition, each marked by a bedding plane.

bedding plane: an ancient surface on which sediment built up. Sedimentary rocks often split along bedding planes.

biotite: a black-coloured form of mica.

body wave: a seismic wave that can travel through the interior of the earth. P waves and S waves are body waves.

boss: an upwards extension of a batholith. A boss may once have been a magma chamber.

botryoidal: the shape of a mineral that resembles a bunch of grapes, e.g. haematite the crystals of which are often arranged in massive clumps, giving a surface covered with spherical bulges.

butte: a small mesa.

calcareous: composed mainly of calcium carbonate.

calcite: a mineral composed of calcium carbonate.

caldera: the collapsed cone of a volcano. It sometimes contains a crater lake.

Caledonian Mountain-Building Period, Caledonian Orogeny: a major mountain-building period in the Lower Paleozoic Era that reached its climax at the end of the Silurian Period (430 to 395 million years ago). An early phase affected only North America and made part of the Appalachian Mountain Belt.

Cambrian, Cambrian Period: the first period of geological time in the Paleozoic Era, beginning 570 million years ago and ending 500 million years ago.

carbonate minerals: minerals formed with carbonate ions (e.g. calcite).

Carboniferous, Carboniferous Period: a period of geological time between about 345 and 280 million years ago. It is often divided into the Early Carboniferous Epoch (345 to 320 million years ago) and the Late Carboniferous Epoch (320 to 280 million years ago). The Late Carboniferous is characterised by large coal-forming swamps. In North America the Carboniferous is usually divided into the Mississippian (= Lower Carboniferous) and Pennsylvanian (= Upper Carboniferous) periods.

cast, fossil: the natural filling of a mould by sediment or minerals that were left when a fossil dissolved after being enclosed by rock.

Cenozoic, Cenozoic Era: the most recent era of geological time, beginning 65 million years ago and continuing to the present.

central vent volcano: *see* stratovolcano.

chemical compound: a substance made from the chemical combination of two or more elements.

chemical rock: a rock produced by chemical precipitation (e.g. halite).

chemical weathering: the decay of a rock through the chemical action of water containing dissolved acidic gases.

cinder cone: a volcanic cone made entirely of cinders. Cinder cones have very steep sides.

class: the level of biological classification below a phylum.

clast: an individual grain of a rock.

clastic rock: a sedimentary rock that is made up of fragments of pre-existing rocks, carried by gravity, water, or wind (e.g. conglomerate, sandstone).

cleavage: the tendency of some minerals to break along one or more smooth surfaces.

coal: the carbon-rich, solid mineral derived from fossilised plant remains. Found in sedimentary rocks. Types of coal include bituminous, brown, lignite, and anthracite. A fossil fuel.

complex volcano: a volcano that has had an eruptive history and which produces two or more vents.

composite volcano: *see* stratovolcano.

concordant coast: a coast where the geological structure is parallel to the coastline. *See also* discordant coastline.

conduction (of heat): the transfer of heat between touching objects.

conglomerate: a coarse-grained sedimentary rock with grains larger than 2 mm.

contact metamorphism: metamorphism that occurs owing to direct contact with a molten magma. *See also* regional metamorphism.

continental drift: the theory suggested by Alfred Wegener that earth's continents were originally one land mass which split up to form the arrangement of continents we see today.

continental shelf: the ocean floor from the coastal shore of continents to the continental slope.

continental shield: the ancient and stable core of a tectonic plate. Also called a shield.

convection: the slow overturning of a liquid or gas that is heated from below.

cordillera: a long mountain belt consisting of many mountain ranges.

core: the innermost part of the earth. The earth's core is very dense, rich in iron, partly molten, and the source of the earth's magnetic field. The inner core is solid and has a radius of about 1300 kilometres. The outer core is fluid and is about 2100 kilometres thick. S waves cannot travel through the outer core.

cracking: the breaking up of a hydrocarbon compound into simpler constituents by means of heat.

crater lake: a lake found inside a caldera.

craton: *see* shield.

Cretaceous, Cretaceous Period: the third period of the Mesozoic Era. It lasted from about 135 to 65 million years ago. It was a time of chalk formation and when many dinosaurs lived.

cross-bedding: a pattern of deposits in a sedimentary rock in which many thin layers lie at an angle to the bedding planes, showing that the sediment was deposited by a moving fluid. Wind-deposited cross-beds are often bigger than water-deposited beds.

crust: the outermost layer of the earth, typically 5 km under the oceans and 50 to 100 km thick under continents. It makes up less than 1 per cent of the earth's volume.

crustal plate: *see* tectonic plate.

crystal: a mineral that has a regular geometric shape and is bounded by smooth, flat faces.

crystal system: a group of crystals with the same arrangement of axes.

crystalline: a mineral that has solidified but has been unable to produce well-formed crystals. Quartz and halite are commonly found as crystalline masses.

crystallisation: the formation of crystals.

cubic: a crystal system in which crystals have 3 axes all at right angles to one another and of equal length.

cuesta: a ridge in the landscape formed by a resistant band of dipping rock. A cuesta has a steep scarp slope and a more gentle dip slope.

current bedding: a pattern of deposits in a sedimentary rock in which many thin layers lie at an angle to the bedding planes, showing that the sediment was deposited by a current of water.

cyclothem: a repeating sequence of rocks found in coal strata.

delta: a triangle of deposition produced where a river enters a sea or lake.

deposit, deposition: the process of laying down material that has been transported in suspension or solution by water, ice, or wind. A deposit is the material laid down by deposition (e.g. salt deposits).

destructive plate boundary: a line where plates collide and where one plate is subducted into the mantle.

Devonian, Devonian Period: the fourth period of geological time in the Palaeozoic Era, from 395 to 345 million years ago.

diorite: an igneous plutonic rock between gabbro and granite; the plutonic equivalent of andesite.

dip: the angle that a bedding plane or fault makes with the horizontal.

dip slope: the more gently sloping part of a cuesta whose surface often parallels the dip of the strata.

discontinuity: a gap in deposition, perhaps caused by the area being lifted above the sea so that erosion, rather than deposition, occurred for a time.

discordant coast: a coast where the rock structure is at an angle to the line of the coast. *See also* concordant coastline.

displacement: the distance over which one piece of rock is pushed relative to another.

dissolve: to break down a substance into a solution without causing a reaction.

distillation: the boiling off of volatile materials, leaving a residue.

dolomite: a mineral composed of calcium magnesium carbonate.

dome: a circular, uplifted region of rocks taking the shape of a dome and found in some areas of folded rocks. Rising plugs of salt will also dome up the rocks above them. They sometimes make oil traps.

dormant volcano: a volcano that shows no signs of activity but which has been active in the recent past.

drift: a tunnel drilled in rock and designed to provide a sloping route for carrying out ore or coal by means of a conveyor belt.

dyke: a wall-like sheet of igneous rock that cuts across the layers of the surrounding rocks.

dyke swarm: a collection of hundreds or thousands of parallel dykes.

earthquake: shaking of the earth's surface caused by a sudden movement of rock within the earth.

element: a fundamental chemical building block. A substance that cannot be separated into simpler substances by any chemical means. Oxygen and sulphur are examples of elements.

eon: the largest division of geological time. An eon is subdivided into eras. Precambrian time is divided into the Archean (earlier than 2.5 billion years ago) and Proterozoic eons (more recent than 2.5 billion years ago). The Phanerozoic Eon includes the Cambrian Period to the present.

epicentre: the point on the earth's surface directly above the focus (hypocentre) of an earthquake.

epoch: a subdivision of a geological period in the geological time scale (e.g. Pleistocene Epoch).

era: a subdivision of a geological eon in the geological time scale (e.g. Cenozoic Era). An era is subdivided into periods.

erode, erosion: the twin processes of breaking down a rock (called weathering) and then removing the debris (called transporting).

escarpment: the crest of a ridge made of dipping rocks.

essential mineral: the dominant mineral constituents of a rock used to classify it.

evaporite: a mineral or rock formed as the result of evaporation of salt-laden water, such as a lagoon or salt lake.

exoskeleton: another word for shell. Applies to invertebrates.

extinct volcano: a volcano that has shown no signs of activity in historic times.

extrusive rock, extrusion: an igneous volcanic rock that has solidified on the surface of the earth.

facet: the cleaved face of a mineral. Used in describing jewellery.

facies: physical, chemical, or biological variations in a sedimentary bed of the same geological age (e.g. sandy facies, limestone facies).

family: a part of the classification of living things above a genus.

fault: a deep fracture or zone of fractures in rocks along which there has been displacement of one side relative to the other. It represents a weak point in the crust and upper mantle.

fault scarp: a long, straight, steep slope in the landscape that has been produced by faulting.

feldspar: the most common silicate mineral. It consists of two forms: plagioclase and orthoclase.

ferromagnesian mineral: dark-coloured minerals, such as augite and hornblende, which contain relatively high proportions of iron and magnesium and low proportions of silica.

fissure: a substantial crack in a rock.

fjord: a glaciated valley in a mountainous area coastal area that has been partly flooded by the sea.

focal depth: the depth of an earthquake focus below the surface.

focus: the origin of an earthquake, directly below the epicentre.

fold: arched or curved rock strata.

fold axis: line following the highest arching in an anticline, or the lowest arching in a syncline.

fold belt: a part of a mountain system containing folded sedimentary rocks.

foliation: a texture of a rock (usually schist) that resembles the pages in a book.

formation: a word used to describe a collection of related rock layers or beds. A number of related beds make a member; a collection of related members makes up a formation. Formations are often given location names, e.g. Toroweap Formation, the members of which are a collection of dominantly limestone beds.

fossil: any evidence of past life, including remains, traces and imprints.

fossil fuel: any fuel that was formed in the geological past from the remains of living organisms. The main fossil fuels are coal and petroleum (oil and natural gas).

fraction: one of the components of crude oil that can be separated from others by heating and then by cooling the vapour.

fracture: a substantial break across a rock.

fracture zone: a region in which fractures are common. Fracture zones are particularly common in folded rock and near faults.

frost shattering: the process of breaking pieces of rock through the action of freezing and melting of rainwater

gabbro: alkaline igneous plutonic rock, typically showing dark-coloured crystals; plutonic equivalent of basalt.

gallery: a horizontal access tunnel in a mine.

gangue: the unwanted mineral matter found in association with a metal.

gem: a mineral, usually in crystal form, that is regarded as having particular beauty and value.

genus: (*pl.* genera) the biological classification for a group of closely related species.

geode: a hollow lump of rock (nodule) that often contains crystals.

geological column: a columnar diagram showing the divisions of geological time (eons, eras, periods, and epochs).

geological eon: *see* eon.

geological epoch: *see* epoch.

geological era: *see* era.

geological period: a subdivision of a geological era (e.g. Carboniferous Period). A period is subdivided into epochs.

geological system: a term for an accumulation of strata that occurs during a geological period (e.g. the Ordovician System is the rocks deposited during the Ordovician Period). Systems are divided into series.

geological time: the history of the earth revealed by its rocks.

geological time scale: the division of geological time into eons, era, periods, and epochs.

geosyncline: a large, slowly subsiding region marginal to a continent where huge amounts of sediment accumulate. The rocks in a geosyncline are eventually lifted to form mountain belts.

gneiss: a metamorphic rock showing large grains.

graben: a fallen block of the earth's crust forming a long trough, separated on all sides by faults. Associated with rift valleys.

grain: a particle of a rock or mineral.

granite: an acidic, igneous, plutonic rock containing free quartz, typically light in colour; plutonic equivalent of rhyolite.

grit: grains larger than sand but smaller than stones.

groundmass: *see* matrix.

group: a word used to describe a collection of related rock layers, or beds. A number of related beds make a member; a collection of related members makes up a formation; a collection of related formations makes a group.

gypsum: a mineral made of calcium sulphate.

halide minerals: a group of minerals (e.g. halite) that contain a halogen element (elements similar to chlorine) bonded with another element. Many are evaporite minerals.

halite: a mineral made of sodium chloride.

Hawaiian-type eruption: a name for a volcanic eruption that mainly consists of lava fountains.

hexagonal: a crystal system in which crystals have 3 axes all at 120 degrees to one another and of equal length.

hogback: a cuesta where the scarp and dip slopes are at about the same angle.

hornblende: a dark-green silicate mineral of the amphibole group containing sodium, potassium, calcium, magnesium, iron and aluminium.

horst: a raised block of the earth's crust separated on all sides by faults. Associated with rift valleys.

hot spot: a place where a fixed mantle magma plume reaches the surface.

hydraulic action: the erosive action of water pressure on rocks.

hydrothermal: a change brought about in a rock or mineral due to the action of superheated mineral-rich fluids, usually water.

hypocentre: the calculated location of the focus of an earthquake.

ice wedging: *see* frost shattering.

Icelandic-type eruption: a name given to a fissure type of eruption.

igneous rock: rock formed by the solidification of magma. Igneous rocks include volcanic and plutonic rocks.

impermeable: a rock that will not allow a liquid to pass through it.

imprint: a cast left by a former life form.

impurities: small amounts of elements or compounds in an otherwise homogeneous mineral.

index fossil: a fossil used as a marker for a particular part of geological time.

intrusive rock, intrusion: rocks that have formed from cooling magma below the surface. When inserted amongst other rocks, intruded rocks are called an intrusion.

invertebrate: an animal with an external skeleton.

ion: a charged particle.

island arc: a pattern of volcanic islands that follows the shape of an arc when seen from above.

isostacy: the principle that a body can float in a more dense fluid. The same as buoyancy, but used for continents.

joint: a significant crack between blocks of rock, normally used in the context of patterns of cracks.

Jurassic, Jurassic Period: the second geological period in the Mesozoic Era, lasting from about 190 to 135 million years ago.

kingdom: the broadest division in the biological classification of living things.

laccolith: a lens-shaped body of intrusive igneous rock with a dome-shaped upper surface and a flat bottom surface.

landform: a recognisable shape of part of the landscape, for example, a cuesta.

landslide: the rapid movement of a slab of soil down a steep hillslope.

lateral fault: *see* thrust fault.

laterite: a surface deposit containing a high proportion of iron.

lava: molten rock material extruded onto the surface of the earth.

lava bomb: *see* volcanic bomb.

law of superposition: the principle that younger rock is deposited on older.

limestone: a carbonate sedimentary rock composed of more than half calcium carbonate.

lithosphere: that part of the crust and upper mantle which is brittle and makes up the tectonic plates.

lode: a mining term for a rock containing many rich ore-bearing minerals. Similar to vein.

Love wave, L wave: a major type of surface earthquake wave that shakes the ground surface at right angles to the direction in which the wave is travelling. It is named after A.E.H. Love, the English mathematician who discovered it.

lustre: the way in which a mineral reflects light. Used as a test when identifying minerals.

magma: the molten material that comes from the mantle and which cools to form igneous rocks.

magma chamber: a large cavity melted in the earth's crust and filled with magma. Many magma chambers are plumes of magma that have melted their way from the mantle to the upper part of the crust. When a magma chamber is no longer supplied with molten magma, the magma solidifies to form a granite batholith.

mantle: the layer of the earth between the crust and the core. It is approximately 2900 kilometres thick and is the largest of the earth's major layers.

marginal accretion: the growth of mountain belts on the edges of a shield.

mass extinction: a time when the majority of species on the planet were killed off.

matrix: the rock or sediment in which a fossil is embedded; the fine-grained rock in which larger particles are embedded, for example, in a conglomerate.

mechanical weathering: the disintegration of a rock by frost shattering/ice wedging.

mesa: a large detached piece of a tableland.

Mesozoic, Mesozoic Era: the geological era between the Palaeozoic and the Cenozoic eras. It lasted from about 225 to 65 million years ago.

metamorphic aureole: the region of contact metamorphic rock that surrounds a batholith.

metamorphic rock: any rock (e.g. schist, gneiss) that was formed from a pre-existing rock by heat and pressure.

meteorite: a substantial chunk of rock in space.

micas: a group of soft, sheet-like silicate minerals (e.g. biotite, muscovite).

mid-ocean ridge: a long mountain chain on the ocean floor where basalt periodically erupts, forming new oceanic crust.

mineral: a naturally occurring inorganic substance of definite chemical composition (e.g. calcite, calcium carbonate).

More generally, any resource extracted from the ground by mining (includes metal ores, coal, oil, gas, rocks, etc.).

mineral environment: the place where a mineral or a group of associated minerals form. Mineral environments include igneous, sedimentary, and metamorphic rocks.

mineralisation: the formation of minerals within a rock.

Modified Mercalli Scale: a scale for measuring the impact of an earthquake. It is composed of 12 increasing levels of intensity, which range from imperceptible, designated by Roman numeral I, to catastrophic destruction, designated by XII.

Mohorovicic discontinuity: the boundary surface that separates the earth's crust from the underlying mantle. Named after Andrija Mohorovicic, a Croatian seismologist.

Mohs' Scale of Hardness: a relative scale developed to put minerals into an order. The hardest is 10 (diamond), and the softest is 1 (talc).

monoclinic: a crystal system in which crystals have 2 axes all at right angles to one another, and each axis is of unequal length.

mould: an impression in a rock of the outside of an organism.

mountain belt: a region where there are many ranges of mountains. The term is often applied to a wide belt of mountains produced during mountain building.

mountain building: the creation of mountains as a result of the collision of tectonic plates. Long belts or chains of mountains can form along the edge of a continent during this process. Mountain building is also called orogeny.

mountain building period: a period during which a geosyncline is compressed into fold mountains by the collision of two tectonic plates. Also known as orogenesis.

mudstone: a fine-grained, massive rock formed by the compaction of mud.

nappe: a piece of a fold that has become detached from its roots during intensive mountain building.

native metal: a metal that occurs uncombined with any other element.

natural gas: *see* petroleum.

normal fault: a fault in which one block has slipped down the face of another. It is the most common kind of fault and results from tension.

nueé ardente: another word for pyroclastic flow.

ocean trench: a deep, steep-sided trough in the ocean floor caused by the subduction of oceanic crust beneath either other oceanic crust or continental crust.

olivine: the name of a group of magnesium iron silicate minerals that have an olive colour.

order: a level of biological classification between class and family.

Ordovician, Ordovician Period: the second period of geological time within the Palaeozoic Era. It lasted from about 500 to 430 million years ago.

ore: a rock containing enough useful metal or fuel to be worth mining.

ore mineral: a mineral that occurs in sufficient quantity to be mined for its metal. The compound must also be easy to process.

organic rocks: rocks formed by living things, for example, coal.

orthoclase: the form of feldspar that is often pink in colour and which contains potassium as important ions.

orogenic belt: a mountain belt.

orogeny: a period of mountain building. Orogenesis is the process of mountain building and the creation of orogenic belts.

orthorhombic: a crystal system in which crystals have 3 axes all at right angles to one another but of unequal length.

outcrop: the exposure of a rock at the surface of the earth.

overburden: the unwanted layer(s) of rock above an ore or coal body.

oxide minerals: a group of minerals in which oxygen is a major constituent. A compound in which oxygen is bonded to another element or group.

Pacific Ring of Fire: the ring of volcanoes and volcanic activity that circles the Pacific Ocean. Created by the collision of the Pacific Plate with its neighbouring plates.

pahoehoe lava: the name for a form of lava that has a smooth surface.

Palaeozoic, Palaeozoic Era: a major interval of geological time. The Palaeozoic is the oldest era in which fossil life is commonly found. It lasted from about 570 to 225 million years ago.

palaeomagnetism: the natural magnetic traces that reveal the intensity and direction of the earth's magnetic field in the geological past.

pegmatite: an igneous rock (e.g. a dyke) of extremely coarse crystals.

Pelean-type eruption: a violent explosion dominated by pyroclastic flows.

period: *see* geological period.

permeable rock: a rock that will allow a fluid to pass through it.

Permian, Permian Period: the last period of the Palaeozoic Era, lasting from about 280 to 225 million years ago.

petrified: when the tissues of a dead plant or animal have been replaced by minerals, such as silica, they are said to be petrified (e.g. petrified wood).

petrified forest: a large number of fossil trees. Most petrified trees are replaced by silica.

petroleum: the carbon-rich, and mostly liquid, mixture produced by the burial and partial alteration of animal and plant remains. Petroleum is found in many sedimentary rocks. The liquid part of petroleum is called oil, the gaseous part is known as natural gas. Petroleum is an important fossil fuel.

petroleum field: a region from which petroleum can be recovered.

Phanerozoic Eon: the most recent eon, beginning at the Cambrian Period, some 570 million years ago, and extending up to the present.

phenocryst: an especially large crystal (in a porphyritic rock), embedded in smaller mineral grains.

phylum: (*pl.* phyla) biological classification for one of the major divisions of animal life and second in complexity to kingdom. The plant kingdom is not divided into phyla but into divisions.

placer deposit: a sediment containing heavy metal grains (e.g. gold) that have weathered out of the bedrock and are concentrated on a stream bed or along a coast.

plagioclase: the form of feldspar that is often white or grey and which contains sodium and calcium as important ions.

planetismals: small embryo planets.

plate: *see* tectonic plate.

plateau: an extensive area of raised flat land. The cliff-like edges of a plateau may, when eroded, leave isolated features such as mesas and buttes. *See also* tableland.

plate tectonics: the theory that the earth's crust and upper mantle (the lithosphere) are broken into a number of more or less rigid, but constantly moving, slabs or plates.

Plinian-type eruption: an explosive eruption that sends a column of ash high into the air.

plug: *see* volcanic plug

plunging fold: a fold whose axis dips, or plunges, into the ground.

plutonic rock: an igneous rock that has solidified at great depth and contains large crystals due to the slowness of cooling (e.g. granite, gabbro).

porphyry, porphyritic rock: an igneous rock in which larger crystals (phenocrysts) are enclosed in a fine-grained matrix.

Precambrian, Precambrian time: the whole of earth history before the Cambrian Period. Also called Precambrian Era and Precambrian Eon.

precipitate: a substance that has settled out of a liquid as a result of a chemical reaction between two chemicals in the liquid.

Primary Era: an older name for the Palaeozoic Era.

prismatic: a word used to describe a mineral that has formed with one axis very much longer than the others.

Proterozoic Eon: *see* eon.

P wave, primary wave, primary seismic wave: P waves are the fastest body waves. The waves carry energy in the same line as the direction of the wave. P waves can travel through all layers of the earth and are generally felt as a thump. *See also* S wave.

pyrite: iron sulphide. It is common in sedimentary rocks that were poor in oxygen, and sometimes forms fossil casts.

pyroclastic flow: solid material ejected from a volcano, combined with searingly hot gases, which together behave as a high-density fluid. Pyroclastic flows can do immense damage, as was the case with Mount Saint Helens.

pyroclastic material: any solid material ejected from a volcano.

Quaternary, Quaternary Period: the second period in the Cenozoic Era, beginning about 1.6 million years ago and continuing to the present day.

radiation: the transfer of energy between objects that are not in contact.

radioactive dating: the dating of a material by the use of its radioactive elements. The rate of decay of any element changes in a predictable way, allowing a precise date to be given of when the material was formed.

rank: a name used to describe the grade of coal in terms of its possible heat output. The higher the rank, the more the heat output.

Rayleigh wave: a type of surface wave having an elliptical motion similar to the waves caused when a stone is dropped into a pond. It is the slowest, but often the largest and most destructive, of the wave types caused by an earthquake. It is usually felt as a rolling or rocking motion and, in the case of major earthquakes, can be seen as they approach. Named after Lord Rayleigh, the English physicist who predicted its existence.

regional metamorphism: metamorphism resulting from both heat and pressure. It is usually connected with mountain building and occurs over a large area. *See also* contact metamorphism.

reniform: a kidney-shaped mineral habit (e.g. hematite).

reservoir rock: a permeable rock in which petroleum accumulates.

reversed fault: a fault where one slab of the earth's crust rides up over another. Reversed faults are only common during plate collision.

rhyolite: acid, igneous, volcanic rock, typically light in colour; volcanic equivalent of granite.

ria: the name for a partly flooded coastal river valley in an area where the landscape is hilly.

Richter Scale: the system used to measure the strength of an earthquake. Developed by Charles Richter, an American, in 1935.

rift, rift valley: long troughs on continents and mid-ocean ridges that are bounded by normal faults.

rifting: the process of crustal stretching that causes blocks of crust to subside, creating rift valleys.

rock: a naturally occurring solid material containing one or more minerals.

rock cycle: the continuous sequence of events that cause mountains to be formed, then eroded, before being formed again.

rupture: the place over which an earthquake causes rocks to move against one another.

salt dome: a balloon-shaped mass of salt produced by salt being forced upwards under pressure.

sandstone: a sedimentary rock composed of cemented sand-sized grains 0.06–2 mm in diameter.

scarp slope: the steep slope of a cuesta.

schist: a metamorphic rock characterised by a shiny surface of mica crystals all orientated in the same direction.

scoria: the rough, often foam-like rock that forms on the surface of some lavas.

seamount: a volcano that rises from the sea bed.

Secondary Era: an older term for a geological era. Now replaced by Mesozoic Era.

sediment: any solid material that has settled out of suspension in a liquid.

sedimentary rock: a layered clastic rock formed through the deposition of pieces of mineral, rock, animal or vegetable matter.

segregation: the separation of minerals.

seismic gap: a part of an active fault where there have been no earthquakes in recent times.

seismic wave: a wave generated by an earthquake.

series: the rock layers that correspond to an epoch of time.

shadow zone: the region of the earth that experiences no shocks after an earthquake.

shaft: a vertical tunnel that provides access or ventilation to a mine.

shale: a fine-grained sedimentary rock made of clay minerals with particle sizes smaller than 2 microns.

shield: the ancient and stable core of a tectonic plate. Also called a continental shield.

shield volcano: a volcano with a broad, low-angled cone made entirely of lava.

silica, silicate: silica is silicon dioxide. It is a very common mineral, occurring as quartz, chalcedony, etc. A silicate is any mineral that contains silica.

sill: a tabular, sheet-like body of intrusive igneous rock that has been injected between layers of sedimentary or metamorphic rock.

Silurian, Silurian Period: the name of the third geological period of the Palaeozoic Era. It began about 430 and ended about 395 million years ago.

skarn: a mineral deposit formed by the chemical reaction of hot acidic fluids and carbonate rocks.

slag: waste rock material that becomes separated from the metal during smelting.

slate: a low-grade metamorphic rock produced by pressure, in which the clay minerals have arranged themselves parallel to one another.

slaty cleavage: a characteristic pattern found in slates in which the parallel arrangement of clay minerals causes the rock to fracture (cleave) in sheets.

species: a population of animals or plants capable of interbreeding.

spreading boundary: a line where two plates are being pulled away from each other. New crust is formed as molten rock is forced upwards into the gap.

stock: a vertical protrusion of a batholith that pushes up closer to the surface.

stratigraphy: the study of the earth's rocks in the context of their history and conditions of formation.

stratovolcano: a tall volcanic mountain made of alternating layers, or strata, of ash and lava.

stratum: (*pl.* strata) a layer of sedimentary rock.

streak: the colour of the powder of a mineral produced by rubbing the mineral against a piece of unglazed, white porcelain. Used as a test when identifying minerals.

striation: minute parallel grooves on crystal faces.

strike, direction of: the direction of a bedding plane or fault at right angles to the dip.

Strombolian-type eruption: a kind of volcanic eruption that is explosive enough to send out some volcanic bombs.

subduction: the process of one tectonic plate descending beneath another.

subduction zone: the part of the earth's surface along which one tectonic plate descends into the mantle. It is often shaped in the form of an number of arcs.

sulphides: a group of important ore minerals (e.g. pyrite, galena, and sphalerite) in which sulphur combines with one or more metals.

surface wave: any one of a number of waves such as Love waves or Rayleigh waves that shake the ground surface just after an earthquake. *See also* Love waves and Rayleigh waves.

suture: the junction of 2 or more parts of a skeleton; in cephalopods the junction of a septum with the inner surface of the shell wall. It is very distinctive in ammonoids and used to identify them.

S wave, shear or secondary seismic wave: this kind of wave carries energy through the earth like a rope being shaken. S waves cannot travel through the outer core of the earth because they cannot pass through fluids. *See also* P wave.

syncline: a downfold of rock layers in which the rocks slope up from the bottom of the fold. *See also* anticline.

system: *see* geological system.

tableland: another word for a plateau. *See* plateau.

tectonic plate: one of the great slabs, or plates, of the lithosphere (the earth's crust and part of the earth's upper mantle) that covers the whole of the earth's surface. The earth's plates are separated by zones of volcanic and earthquake activity.

Tertiary, Tertiary Period: the first period of the Cenozoic Era. It began 665 and ended about 1.6 million years ago.

thrust fault: *see* reversed fault.

transcurrent fault: *see* lateral fault.

transform fault: *see* lateral fault.

translucent: a description of a mineral that allows light to penetrate but not pass through.

transparent: a description of a mineral that allows light to pass right through.

trellis drainage pattern: a river drainage system where the trunk river and its tributaries tend to meet at right angles.

trench: *see* ocean trench.

Triassic, Triassic Period: the first period of the Mesozoic era. It lasted from about 225 to 190 million years ago.

triclinic: a crystal system in which crystals have 3 axes, none at right angles or of equal length to one another.

tsunami: a very large wave produced by an underwater earthquake.

tuff: a rock made from volcanic ash.

unconformity: any interruption in the depositional sequence of sedimentary rocks.

valve: in bivalves and brachiopods, one of the separate parts of the shell.

vein: a sheet-like body of mineral matter (e.g. quartz) that cuts across a rock. Veins are often important sources of valuable minerals. Miners call such important veins lodes.

vent: the vertical pipe that allows the passage of magma through the centre of a volcano.

vertebrate: an animal with an internal skeleton.

vesicle: a small cavity in a volcanic rock originally created by an air bubble trapped in the molten lava.

viscous, viscosity: sticky, stickiness.

volatile: substances that tend to evaporate or boil off of a liquid.

volcanic: anything from, or of, a volcano. Volcanic rocks are igneous rocks that cool as they are released at the earth's surface – including those formed underwater; typically have small crystals due to the rapid cooling, e.g. basalt, andesite and rhyolite.

volcanic bomb: a large piece of magma thrown out of a crater during an eruption, which solidifies as it travels through cool air.

volcanic eruption: an ejection of ash or lava from a volcano.

volcanic glass: lava that has solidified very quickly and has not had time to develop any crystals. Obsidian is a volcanic glass.

volcanic plug: the solidified core of an extinct volcano.

Vulcanian-type eruption: an explosive form of eruption without a tall ash column or pyroclastic flow.

water gap: a gap cut by a superimposed river, which is still occupied by the river.

weather, weathered, weathering: the process of weathering is the mechanical action of ice and the chemical action of rainwater on rock, breaking it down into small pieces that can then be carried away. *See also* chemical weathering and mechanical weathering.

wind gap: a gap cut by a superimposed river, which is no longer occupied by the river.

Set Index

USING THE SET INDEX

This index covers all eight volumes in the *Earth Science* set:

Volume number	Title
1:	Minerals
2:	Rocks
3:	Fossils
4:	Earthquakes and volcanoes
5:	Plate tectonics
6:	Landforms
7:	Geological time
8:	The earth's resources

An example entry:

Index entries are listed alphabetically.

plagioclase feldspar **1:** *51*; **2:** 10 *see also* feldspars

Volume numbers are in bold and are followed by page references. Articles on a subject are shown by italic page numbers.

In the example above, 'plagioclase feldspar' appears in Volume 1: Minerals on page 51 as a full article and in Volume 2: Rocks on page 10. Many terms also are covered in the GLOSSARY on pages 60–65.

The *see also* refers to another entry where there will be additional relevant information.

D

E